Advanced Foams for Process Intensification

Edited by Chen Yang, Yixiong Lin, Koyo Norinaga and Ting Qiu

Published in London, United Kingdom

Advanced Foams for Process Intensification
http://dx.doi.org/10.5772/intechopen.1009700
Edited by Chen Yang, Yixiong Lin, Koyo Norinaga and Ting Qiu

Contributors
Antonio Montes, Chen Yang, Chen Yang, Chunyang Wang, Clara Pereyra, Diego Valor, Ignacio García-Casas, Jiazhou Li, Jie Niu, Koyo Norinaga, Ludisbel León-Marcos, Moghtada Mobedi, Rongchi Dai, Ruming Pan, Shijie Wu, Shiqi You, Suchada Sirisomboonchai, Ting Qiu, Wei Zhang, Xinyi Chen, Yaning Zhang, Yibo Wu, Yixiong Lin, Yixiong Lin, Yiyu Chen, Youwei Yang, Yuhan Jin, Yuming Zhang, Yun Sun, Zhewen Chen, Zhongmin Wan

First published in London, United Kingdom, 2026 by IntechOpen
IntechOpen is the global imprint of INTECHOPEN LIMITED, registered in England and Wales, registration number: 11086078, 167-169 Great Portland Street, London, W1W 5PF, United Kingdom

For EU product safety concerns: IN TECH d.o.o., Prolaz Marije Krucifikse Kozulić 3, 51000 Rijeka, Croatia, info@intechopen.com or visit our website at intechopen.com.

British Library Cataloguing-in-Publication Data
A catalogue record for this book is available from the British Library

Advanced Foams for Process Intensification
Edited by Chen Yang, Yixiong Lin, Koyo Norinaga and Ting Qiu
p. cm.
Print ISBN 978-1-83634-176-5
Online ISBN 978-1-83634-175-8
eBook (PDF) ISBN 978-1-83634-177-2

If disposing of this product, please recycle the paper responsibly.

Meet the editors

Chen Yang is a Professor at the College of Chemical Engineering, Fuzhou University, China. He received his Ph.D. in 2012 from the University of Shizuoka, Japan. From 2012 to 2014, he worked as a postdoc at the National Polytechnic Institute of Toulouse (INPT), France. He joined Fuzhou University, China, as an Associate Professor in 2014 and was promoted to Full Professor in 2016. His research covers various areas, including transport theory in porous media and chemical process intensification. He has published more than 100 journal papers. Due to the achievement of catalytic packing design, Prof. Yang received several awards, including first prize for the S&T Progress Award (2016, China Petroleum and Chemical Industry Association; 2018, People's Government of Fujian Province).

Yixiong Lin serves as an Associate Professor at the College of Chemical Engineering, Fuzhou University, China. He earned his Ph.D. from Nagoya University, Japan, in 2022. Dr. Lin joined Fuzhou University as an Associate Professor in 2023. His research primarily focuses on CO2 utilization, transport theory in porous media, and chemical process intensification. He has published more than forty journal papers.

Koyo Norinaga is a Professor at the College of Engineering, Nagoya University, Japan. He obtained his Ph.D. from Hokkaido University, Japan, in 1999. From 1999 to 2002, he served as a Research Associate at Tohoku University, Japan. Following this, he was awarded an Alexander von Humboldt Fellowship and conducted research at the Karlsruhe Institute of Technology (KIT) in Germany from 2002 to 2004. He subsequently continued his work as a Senior Scientist at KIT until 2006. In 2009, Prof. Norinaga joined Kyushu University as an Associate Professor. In 2017, he was promoted to Full Professor at Nagoya University. His research primarily focuses on CO_2 capture and utilization, carbon resources, energy, and material-efficient conversion.

Ting Qiu is a Professor and Doctoral Supervisor in Chemical Engineering at Fuzhou University. He earned his Ph.D. from Tianjin University (2003) and specializes in green catalyst design, chemical process intensification, and ultra-high-purity electronic chemical purification for integrated circuits. As principal investigator, he has led over 100 research projects and published over 200 journal papers. Moreover, Prof. Qiu received several awards, including first prize for the S&T Progress Award (2016, China Petroleum and Chemical Industry Association; 2018, People's Government of Fujian Province).

Contents

Preface

The development of sustainable energy solutions and advanced material technologies has assumed unprecedented urgency in response to escalating global climate challenges and rising energy demands. This book, *Advanced Foams for Process Intensification*, provides an interdisciplinary study of functional foam materials and explores the crucial roles they play in process intensification, high-efficiency energy conversion, and environmental sustainability. Integrating first-principles modeling, high-fidelity numerical simulations, and systematic experimental characterization, this book elucidates the structure-performance relationship between foam structure and transport process, offering valuable insights for researchers, engineers, and students.

Foam materials, characterized by their well-defined three-dimensional porous architectures, exceptionally high specific surface areas, and precisely tunable physicochemical properties, have demonstrated significant promise across a range of energy and chemical engineering applications, including catalytic reactors, proton-exchange membrane fuel cells, concentrated solar thermal systems, and lean-burn combustion devices. Nevertheless, realizing their full functional potential necessitates a mechanistic, multi-scale understanding that bridges pore-scale transport and reaction phenomena with macroscopic system-level performance metrics. To bridge this critical knowledge gap, this book synergistically combines high-fidelity computational fluid dynamics (CFD) modeling, physics-informed machine learning methodologies, and advanced additive manufacturing techniques to systematically decode the structure-performance relationships for foam materials.

Chapter 1 examines heat and fluid flow in metal foams using both pore-scale and volume-averaged approaches. It delves into local thermal equilibrium and non-equilibrium models, providing a foundation for analyzing thermal management in high-temperature processes.

Chapter 2 advances the discourse by coupling CFD with deep learning to investigate hierarchical pore structures in foam reactors for dry reforming of methane (DRM). This chapter demonstrates how convolutional neural networks (CNNs) can predict velocity and temperature fields with high accuracy, revealing optimal pore configurations that enhance reaction rates while suppressing coke formation. The integration of AI with traditional simulations marks a significant step toward smart reactor design.

Chapter 3 explores the application of metal foam flow fields in proton exchange membrane fuel cells (PEMFCs), emphasizing water management and output performance. Numerical and experimental studies show that hierarchical pore structures improve gas permeability and reduce pressure drop, resulting in a 9.5% increase in peak power density compared to conventional designs.

Chapter 4 expands the scope to electrochemical applications, where foam electrodes made of metals (e.g., nickel, copper) and carbon are evaluated for water splitting and CO_2 reduction. This chapter highlights key fabrication strategies, including electrodeposition and templating, and discusses how foam morphology influences catalytic activity, gas release, and reactant accessibility.

Chapter 5 presents a numerical study of a solar-driven pyrolysis foam reactor, analyzing the effects of operating conditions, such as xenon lamp power and foam porosity, on photothermal performance. The findings reveal that lower porosity and smaller pore sizes can optimize energy conversion rates, contributing to the development of efficient solar thermal systems.

Chapter 6 addresses the challenge of temperature regulation in exothermic reactors, particularly for CO_2 methanation. To suppress hot spots and improve stability, this study employs numerical simulations to examine packed-bed and monolithic reactors and proposes strategies such as catalyst dilution and "CHESS" monolith designs. This chapter highlights the importance of thermal management in achieving high conversion efficiency.

Chapter 7 investigates the combustion characteristics and flame stability of hydrogen-enriched natural gas (HENG) in porous media burners. It identifies critical hydrogen blending ratios that trigger combustion mode transitions and analyzes mechanisms behind flame instability, offering guidelines for safe and efficient residential gas applications.

Chapter 8 introduces the fundamentals of supercritical CO_2-processed polymeric foams, highlighting their role in sustainable manufacturing and functionalization for applications like drug delivery and catalysis. The discussion emphasizes how parameters such as pressure, temperature, and polymer composition influence foam morphology and functionality, setting the stage for advanced applications in process intensification.

Targeted at an interdisciplinary readership, this book serves researchers and educators in chemical engineering, materials science, and sustainable energy systems, as well as industry professionals engaged in the development and deployment of next-generation porous media solutions. We intend that the conceptual frameworks, methodological insights, and case-based analyses presented herein will catalyze cross-sectoral innovation and collaborative progress toward decarbonized, resource-efficient industrial processes.

Chen Yang, Yixiong Lin and Ting Qiu
College of Chemical Engineering,
Fuzhou University,
China

Koyo Norinaga
College of Engineering,
Nagoya University,
Japan

Chapter 1

Heat and Fluid Flow in Metal Foam: Pore Scale and Volume Averaged Methods, Local Thermal Equilibrium and Nonequilibrium Models

Chunyang Wang and Moghtada Mobedi

Abstract

Improving energy utilization efficiency has always been one of the hot topics. Due to advantages of high thermal conductive porous media for heat transfer enhancement, heat and fluid flow in metal foam has received attentions of researchers in recent years. High thermal conductive porous structure, for example metal foam, can be used for heat transfer enhancement in wide applications such as phase change heat exchanger and thermal energy storage. In this chapter, a summary of different kinds of metal foam for enhancement of heat transfer is provided. The governing equations of fluid flow and heat transfer with solid-liquid phase change in porous media, the volume averaged and pore scale methods, and two models as local thermal equilibrium and nonequilibrium are given in detail. Also, the challenges and outlook of heat transfer in metal foam are discussed.

Keywords: metal foam, heat transfer enhancement, pore scale method, volume averaged method, local thermal equilibrium model, local thermal nonequilibrium model

1. Introduction

A porous medium is a composite material including a solid matrix interspersed with void spaces. These voids can be either interconnected, forming open-cell structures, or isolated, forming closed-cell structures. Among various types of porous materials, metal foams have garnered significant attention in both engineering applications and scientific research. This interest stems from their unique three-dimensional interconnected structures, combined with exceptionally high porosity and high thermal conductivity. These structural characteristics enable metal foams to markedly enhance both heat transfer and fluid flow within systems, making them highly attractive for various thermal management applications. As a result, metal foams find wide-ranging applications in heat transfer fields, including but not limited to heat exchanger [1, 2], cooling system [3, 4], and thermal energy storage [5, 6].

In addition to the conventional studies focusing on single-phase heat and fluid flow, there has been a growing research interest in exploring the phase change phenomena associated with metal foams [7, 8]. In particular, the integration of metal foam into phase change materials (PCMs) has been shown to significantly improve the overall thermal performance of PCMs. The presence of the metal foam not only enhances the thermal conductivity of PCM composite but also increases the solid–liquid interfacial area, which facilitates faster heat exchange. As a result, the overall heat transfer rate during both the melting and solidification processes is greatly improved compared to the use of pure PCMs only [9, 10]. At the pore scale, the parameters of the flow such as velocity, pressure, and temperature exhibit significant spatial irregularities due to the complex and tortuous geometry of the porous structure. Moreover, because of the extremely small size and intricate configuration of the flow passages within the pores, it is extremely challenging to perform accurate local field measurements of fluid dynamic behaviors at this scale using experimental techniques. Consequently, theoretical analyses, particularly those based on solving the Navier–Stokes equations and associated energy equations, have been widely employed to study fluid flow and heat transport in porous media, including metal foams.

Generally, there are two kinds methods for analyzing of heat and fluid flow in metal foam as Pore Scale Method (PSM) and Volume Averaged Method (VAM) [11, 12]. The PSM involves direct numerical simulation at the pore level, capturing detailed flow and temperature fields within the actual geometrical structure, while the VAM simplifies the modeling by averaging properties over a Representative Elementary Volume (REV), thereby enabling more computationally efficient simulations at the macroscopic scale. Furthermore, an important conceptual model when analyzing heat and fluid flow in porous media is the local thermal equilibrium (LTE) model or local thermal nonequilibrium (LTNE) model [13, 14]. Under the local thermal equilibrium model, it is considered that the temperature difference between the solid matrix and fluid phase at the pore scale is negligible compared to macroscopic temperature variations, allowing the use of a single energy equation for the entire medium. However, when significant temperature differences exist between the phases, especially during transient processes or in systems with large thermal conductivity disparities between the solid and fluid, the local thermal nonequilibrium model must be adopted, requiring separate energy equations for the solid and fluid phases.

In this chapter, the governing equations for cases of fluid flow and heat transfer with solid-liquid phase change in metal foam are given. The volume averaged and pore scale methods and two models as local thermal equilibrium and nonequilibrium are discussed in detail. Finally, the challenges and outlook of heat transfer in metal foam are also prospected.

2. Porosity and permeability

For heat and fluid flow in metal foam, two important and fundamental parameters as porosity and permeability should be known. These properties play a decisive role in characterizing the transport behavior within the porous structure. The porosity of a metal foam refers to the fraction of the voids or pores volume, as opposed to the solid material. In other words, porosity quantifies the proportion of the volume that is available for fluid flow and heat transport. A higher porosity generally indicates a greater capacity for fluid movement and potential for enhanced thermal performance. The definition of porosity can be presented as follow.

$$\varepsilon = {}^{V_v}\!/_{V_t} \tag{1}$$

where ε is the porosity of the metal foam, V_v and V_t are the volume of the pores and total volume of the metal foam, respectively. And it is a dimensionless number ranging between 0 and 1, or as a percentage.

In addition to porosity, permeability is another essential parameter that describes the ease with which a fluid can go through the porous medium. It reflects the interconnectedness and size distribution of the pores, and it strongly influences the pressure drop and flow resistance encountered during fluid transport. A thorough understanding of both porosity and permeability is necessary for accurately predicting and optimizing heat and mass transfer processes in metal foam applications. For example, based on the studies of Ozgumus et al. [15] and Celik et al. [16], it is assumed that the friction effect is dominant due to the surface of the metal foam. For one dimensional flow, the momentum equation can be given in Eq. (2). In the numerical determination of permeability, a uniform inlet velocity is applied to a representative unit cell. The continuity and momentum equations at the pore scale are solved and the pressure drop across the cell is computed by the given velocity (i.e., corresponding Reynolds number). This procedure is repeated for a range of inlet velocities. By plotting the variation of dimensionless pressure against the pore-scale Reynolds number, the permeability and inertia coefficient can be derived from the resulting relationship.

$$\Pi = A + B\,\mathrm{Re}, \Pi = -\frac{d_c^2}{\langle u\rangle \mu_f}\frac{d\langle p\rangle^f}{dx}, A = \frac{d_c^2}{K}, B = C_F\left(\frac{d_c^2}{K}\right)^{1/2}, \mathrm{Re} = \frac{\rho_f \langle u\rangle d_c}{\mu_f} \tag{2}$$

where $\Pi, A, B$are dimensionless parameters. $C_F$is inertia coefficient and$K$ is the permeability of the metal foam, and d_c is the averaged particle or fiber diameter.

3. Pore scale and volume averaged methods

There are two principal methods that can be employed to determine the temperature distribution and the behaviors of heat transfer and fluid flow through a porous medium, which are the Pore Scale Method and the Volume Averaged Method. Many studies have been done by applying the Pore Scale Method and the Volume Averaged Method for heat transfer in porous media. **Table 1** shows the comparison between Pore Scale Method and Volume Averaged Method. Each method offers unique insights into the mechanisms of heat transfer in porous structures, with their applicability depending largely on the specific scale and the objectives of the analysis.

Figure 1 indicates the macro scale and pore scale domains. PSM is particularly suited for detailed investigations at the microscale, where the intricate structure of the porous medium is explicitly resolved. In this approach, the governing equations such as the Navier–Stokes equations and the energy conservation equations are solved directly at the level of individual pores. The porous medium is treated as a realistic three-dimensional structure, capturing the exact geometrical features of both the solid matrix and the interconnected pore network. As a result, the parameters such as local velocity, pressure, and temperature fields are obtained with high precision, enabling a thorough understanding of microscale transport phenomena. PSM is particularly advantageous for fundamental studies aimed at exploring the detailed mechanisms of heat and mass transfer at the pore level. It is commonly employed in numerical

Feature	Pore scale method	Volume averaged method
Modeling scale	Microscale (individual pores)	Macroscale (representative volume)
Accuracy	Very high	Moderate to high
Computational cost	Very high (requires fine mesh and large memory)	Low to moderate (coarse mesh sufficient)
Application domain	Small domains (limited by computational resources)	Large domains (system-level simulations)
Numerical methods commonly used	Computational Fluid Dynamics (CFD), Lattice Boltzmann Method (LBM)	Finite Volume Method (FVM), Finite Element Method (FEM)
Limitations	Not practical for large or highly complex systems	May lose fine-scale local information

Table 1.
The comparison between pore scale method and volume averaged method.

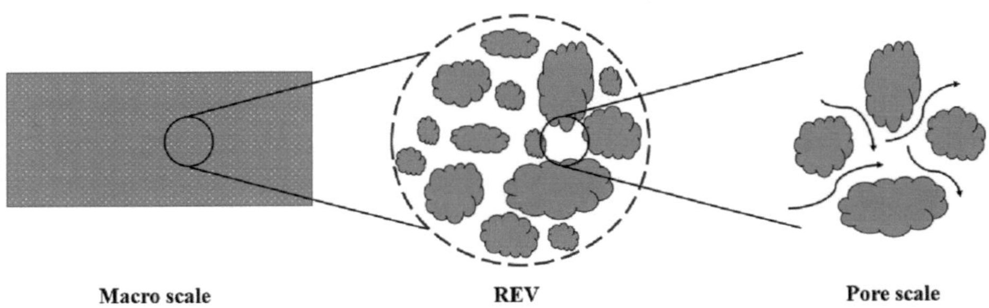

Figure 1.
The macro scale and pore scale domains.

simulations using high-resolution computational techniques, such as CFD or LBM. However, despite its accuracy, PSM faces considerable challenges in practical applications. The complexity and irregularity of realistic porous structures, such as those found in metal foams, necessitate extremely fine spatial discretization and high computational resources. The associated demands for large memory capacity, intensive computation time, and high storage requirements limit the feasibility of PSM for large-scale domains or long-term transient simulations.

To overcome these limitations, the Volume Averaged Method is widely employed, especially in engineering-scale analyses where computational efficiency is critical. VAM simplifies the problem by averaging the microscopic governing equations over a Representative Elementary Volume of the porous medium. This averaging process leads to the formulation of macroscopic transport equations, which describe the overall, or bulk, behavior of the system rather than resolving individual pores. In VAM, distinct solid–fluid interfaces present at the microscale are effectively homogenized, resulting in a continuous medium characterized by effective properties. Critical volume-averaged transport parameters of heat and fluid flow include permeability, effective thermal conductivity, inertia coefficient for both the solid and fluid phases, and the interfacial heat transfer coefficient. And these parameters encapsulate the complex microscale interactions into effective quantities that can be used in macroscopic modeling. The VAM significantly reduces the computational cost compared to PSM, making it feasible to simulate larger domains and longer time periods with reasonable accuracy. As a result, VAM is particularly advantageous for the design and optimization of engineering systems that employ metal foams, such as heat exchangers, thermal energy storage units, and catalytic reactors. It allows engineers and researchers to predict system-level thermal and fluid dynamic performance without the prohibitive computational burden associated with fully resolving the microstructure. In summary, PSM provides a detailed understanding at the pore level for theoretical investigations, while VAM offers a practical and scalable approach for real-world applications where system-level behavior needs to be analyzed efficiently.

3.1 Pore scale governing equations

The continuity and momentum equations of the pore scale analysis are directly solved to obtain the detailed distributions of velocity and pressure for the fluid flowing through the interconnected pore spaces between the solid structures. In parallel, the energy equations for both the solid matrix and the fluid phase are solved independently to determine the temperature distribution throughout the entire computational domain, capturing the complex interactions between heat conduction and convection. At the pore scale, it is critical to explicitly account for the detailed geometry of the porous medium, ensuring that the microstructural effects on transport phenomena are accurately represented. Furthermore, when considering cases where gravitational forces have a non-negligible impact on the flow (e.g., buoyancy-driven flows or natural convection), the governing equations must incorporate the effects of gravity appropriately. The dimensional form of the pore scale governing equations when the gravity also affects the flow can be written as follow [14]. The Eqs. (3)–(6) are continuity, momentum, and energy equations for fluid and solid phases, respectively.

$$\vec{\nabla}\cdot\vec{V}=0 \tag{3}$$

$$\frac{\partial\vec{V}}{\partial t}+\left(\vec{V}.\,\vec{\nabla}\right)\vec{V}=-\frac{1}{\rho_f}\vec{\nabla}p+\nu_f(T)\vec{\nabla}^2.\vec{V}+g\beta\left(T-T_{ref}\right)\vec{j} \tag{4}$$

$$\left(\rho C_p\right)_f\left(\frac{\partial T_f}{\partial t}+\vec{V}.\,\vec{\nabla}T_f\right)=k_f(T)\vec{\nabla}^2T_f \tag{5}$$

$$\left(\rho C_p\right)_s\frac{\partial T_s}{\partial t}=k_s\vec{\nabla}^2T_s \tag{6}$$

where β represents the volumetric thermal expansion coefficient of the working fluid and T_{ref} denotes the reference temperature. The last term in Eq. (4) captures the influence of gravity on the flow when the fluid density varies with temperature. This term is incorporated based on the widely used Boussinesq approximation, which simplifies the treatment of buoyancy effects by assuming that density variations are sufficiently small to affect only the gravitational body force term. Moreover, it is assumed that thermophysical properties of the fluid that are constant, and the fluid is Newtonian and incompressible.

3.2 Volume averaged governing equations

Pore scale analysis can provide highly accurate and detailed results for heat and fluid flow within porous media, as it captures the complex interactions at the microscale level. However, when the physical size or geometric complexity of the porous medium becomes large, the application of pore scale analysis becomes increasingly challenging and computationally expensive. The enormous computational resources required to resolve each pore individually often make this method impractical for large-scale or engineering-level problems. That is why researchers commonly adopt the Volume Averaged Method to analyze heat and fluid flow in porous media across larger domains. The VAM effectively bridges the gap between microscale complexity and macroscale applicability by simplifying the governing equations through an averaging process. In the VAM framework, a scalar or vector quantity is integrated over a representative elementary volume (REV) that contains both phases (solid and fluid). This integration smooths out the microscale discontinuities and allows the definition of continuous fields for the purpose of macroscopic analysis. Thus, the inherently discontinuous nature of the pore structure is homogenized into effective, continuum properties. Since there are two phases as solid and fluid in the porous media, volume averaged and intrinsic volume averaged quantities can be defined as [14]:

$$\langle \varphi \rangle = \frac{1}{V} \int_V \varphi dV \tag{7}$$

$$\langle \varphi \rangle^x = \frac{1}{V_x} \int_{V_x} \varphi dV \tag{8}$$

where $V$and V_x are the total volume and corresponding volume of solid or fluid phase. These volume averaged quantities form the basis for deriving the macroscopic governing equations, which can predict the overall heat and mass transport behavior in the porous medium while substantially reducing the computational burden.

By taking the volume average of the dimensional pore scale governing equations (i.e., Eqs. (3)–(6)), the corresponding dimensional volume averaged governing equations (Eqs. (9)–(12)) can be derived [14]. This averaging process transforms the complex, highly detailed pore-level descriptions into more tractable macroscopic equations that characterize the overall behavior of the porousmedium.

$$\vec{\nabla} \cdot \left\langle \vec{V} \right\rangle = 0 \tag{9}$$

$$\begin{aligned} &\frac{\rho_f}{\varepsilon} \frac{\partial \left\langle \vec{V} \right\rangle^f}{\partial t} + \frac{\rho_f}{\varepsilon^2} \left(\left\langle \vec{V} \right\rangle^f . \vec{\nabla} \right) \left\langle \vec{V} \right\rangle^f = -\vec{\nabla} \langle P \rangle^f + \frac{\mu_f}{\varepsilon} \vec{\nabla}^2 \left\langle \vec{V} \right\rangle^f \\ &- \frac{\mu_f}{K} \left\langle \vec{V} \right\rangle^f - \rho_f \frac{C_F}{\sqrt{K}} \left| \left\langle \vec{V} \right\rangle^f \right| \left\langle \vec{V} \right\rangle^f + \rho_f g \beta (T - T_{ref}) \vec{j} \end{aligned} \tag{10}$$

$$\begin{aligned} &\varepsilon (\rho C_p)_f \frac{\partial \langle T \rangle^f}{\partial t} + (\rho C_P)_f \left\langle \vec{V} \right\rangle^f . \vec{\nabla} \langle T \rangle^f = k_{eff,f}(T) \vec{\nabla}^2 \langle T \rangle^f \\ &+ h_v \left(\langle T \rangle^s - \langle T \rangle^f \right) \end{aligned} \tag{11}$$

$$(1 - \varepsilon)(\rho C_p)_s \frac{\partial \langle T \rangle^s}{\partial t} = k_{eff,s} \vec{\nabla}^2 \langle T \rangle^s - h_v \left(\langle T \rangle^s - \langle T \rangle^f \right) \tag{12}$$

In these equations, ε, K and C_F represent the porosity, permeability, and inertia coefficient of the porous medium, respectively. Each of these parameters plays a crucial role in characterizing the resistance to fluid flow and the overall transport behavior within the porous structure. The third term on the right side of Eq. (10) is the Darcy term, which accounts for the viscous resistance exerted by the porous matrix on the fluid. The fourth term represents the inertial effects that become significant at higher flow velocities and are modeled through a nonlinear Forchheimer correction. The fifth term captures the influence of buoyancy forces arising due to gravity and temperature-induced density variations, and is incorporated using the Boussinesq approximation. The parameter h_v denotes the volumetric interfacial heat transfer coefficient between the solid matrix and fluid, quantifying the rate of thermal energy exchange across the solid–fluid interfaces within the porous matrix. Furthermore, $k_{eff,s}$ and $k_{eff,f}$ are the effective thermal conductivities of the solid and fluid phases, which are modified from their intrinsic values due to the porous structure's geometry and connectivity. The heat transfer effect between the solid and fluid phases are expressed in Eqs. (13) and (14). $k_{eff,s}$ and $k_{eff,f}$can be defined as,

$$k_{eff,s} = (1 - \varepsilon^*)k_s \tag{13}$$
$$k_{eff,f} = \varepsilon^* k_f$$

The ε^* is the effective porosity and it can be found from following equation.

$$k_{eff} = \varepsilon^* k_f + (1 - \varepsilon^*)k_s \tag{14}$$

k_{eff} can be calculated numerically. To determine the effective thermal conductivities, a representative pore-scale cell within an isotropic domain can be selected. A temperature gradient is applied along the x-axis, while the remaining boundaries are thermally insulated. The steady-state heat conduction equation is then solved separately for two kinds of different fluid. Based on the solutions, the effective thermal conductivity can be obtained. Subsequently, Eq. (14) is used to calculate the value of ε^*, and finally, the effective thermal conductivities of the solid phase ($k_{eff,s}$) and fluid phase ($k_{eff,f}$) are easily determined by Eq. (13).

As mentioned before, these macroscopic equations incorporate effective properties such as permeability, inertia coefficient, effective thermal conductivity, and interfacial heat transfer coefficient, which must either be determined experimentally or computed through detailed numerical simulations based on the pore structure. The calculation methods for permeability and inertia coefficient have been extensively studied in the literature, with detailed procedures available in Refs. [15, 16]. These references provide systematic approaches for determining the flow resistance characteristics of porous media based on both experimental and numerical techniques. Wang et al. [11] conducted a comprehensive study on heat transfer in a three-dimensional lattice metal foam structure. In their investigation, the permeability and inertia coefficient were quantitatively evaluated. The study involved solving the continuity and momentum equations at the pore scale for a representative unit cell, subjected to a range of different inlet velocities to simulate various flow conditions. By analyzing the simulation results, the relationship between the dimensionless pressure drop and the pore-scale Reynolds number was established. From the slope and intercept of the resulting curve, the permeability and inertia coefficient of the porous media were determined based on the obtained data.

4. Local thermal equilibrium and no-equilibrium models

There are two primary models that are commonly employed in the analysis of heat transfer within metal foams: local thermal equilibrium and nonequilibrium, as shown in **Figure 2**. Generally, there are many studies on local thermal equilibrium or local thermal nonequilibrium model are done, especially heat transfer with phase change in metal foam. **Table 2** shows some reported studies refer to local thermal equilibrium or nonequilibrium model for phase change in metal foam, and the applied porosity and main findings are also presented. These assumptions play a pivotal role in shaping the formulation of the energy equations and significantly influence the interpretation of heat transfer behavior in porous structures, particularly in systems involving complex interactions between solid and fluid phases.

Under the local thermal equilibrium model, it is presumed that the temperature difference between the solid matrix and the fluid within the pores at the microscopic (pore) scale is negligible when compared to the temperature gradients observed at the macroscopic scale. In other words, it is assumed that the solid and fluid temperatures are essentially identical at any given point in the porous medium. This assumption allows the system to be modeled using a single averaged energy equation that describes the overall temperature distribution across the entire domain, significantly simplifying the modeling process and reducing computational complexity. The LTE model is particularly valid in cases:

- The thermal conductivity of the solid and fluid phases is similar;
- The interfacial heat transfer between the solid and fluid is very efficient;
- The system has reached a steady state or varies slowly with time.

Since the temperature of fluid phase is equal or close to the temperature of solid phase, only one energy equation is enough to express the heat transfer behavior in porous media, and the energy equation can be defined as [25],

$$(\rho C_p)_{eff}\frac{\partial T}{\partial t} + (\rho C_p)_f \vec{V}.\,\vec{\nabla}T = k_{eff}(T)\vec{\nabla}^2 T + q \tag{15}$$

where k_{eff} is effective thermal conductivity and expressed as $\varepsilon^* k_f + (1-\varepsilon^*)k_s$, and q is the heat source term.

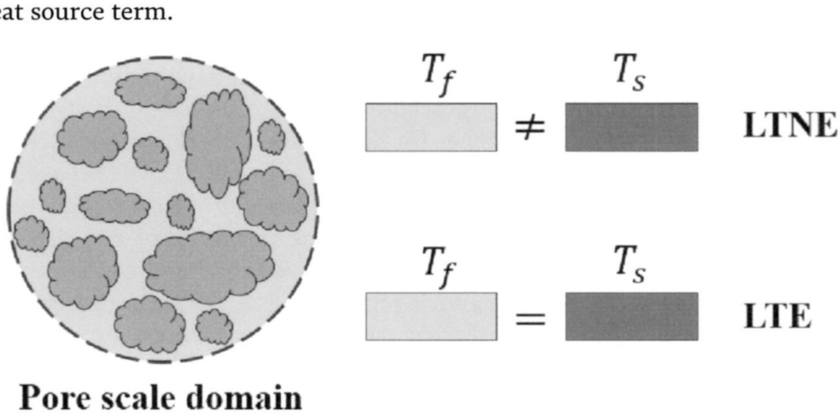

Figure 2.
The local thermal equilibrium and local thermal nonequilibrium models.

Reference	Models	Porosity	Main findings
Zhao et al. [17]	Local thermal equilibrium	0.8–0.95	The porosity for the fully filled scenarios has a great effect on the PCM melting and solidification processes.
Liu et al. [18]	Local thermal nonequilibrium	0.94	Design on graded porosity reduces the complete melting time by 21.1% compared with the non-gradient structure.
Pu et al. [19]	Local thermal nonequilibrium	0.89–0.99	An optimal configuration 0.99–0.97-0.89 of negative gradient was recommended to further reduce the complete melting time by 23.7%.
Shakibi et al. [20]	Local thermal nonequilibrium	0.4, 0.8, 0.9	It was found that average electrical efficiency for the model with a porosity of 0.9, 0.8, 0.4, and no-foam state are 13.805%, 13.794%, 13.761%, and 13.714%, respectively.
Marri and Balaji [21]	Local thermal nonequilibrium	0.9, 0.94, 0.97 and PPI of 8. 14 and 20	PCM heat sink with nonuniform porosity and nonuniform PPI density gradients created with tri-layer metal foams have almost the same performance as a bilayer metal foam with enhancement ratio up to 4 and 4.4 times, respectively, over the baseline case (i.e., PCM heat sink without metal foams).
Briache et al. [22]	Local thermal equilibrium	0.8, 0.9, 0.95	Metal foams with 80% porosity can reduce 64.6% melting time and increase 145% storage rate.
Hu et al. [23]	Local thermal nonequilibrium	0.85, 0.9, 0.95	For the critical temperature of 40°C, the maximum operation time of heat sinks with RT31 is about 5 times greater than that of heat sinks with RT55.
Yang et al. [24]	Local thermal equilibrium	0.9–0.98	The pin fin-metal foam hybrid structure with graded porosity can further improve the solidification rate by 34.7%.

Table 2.
Some studies on local thermal equilibrium or local thermal nonequilibrium model for phase change in metal foam.

However, while the LTE model has been widely adopted in earlier studies due to its simplicity and the ease with which it can be applied, its applicability may be questioned, particularly a common occurrence in thermal systems utilizing metal foams in transient heat transfer processes. In such cases, there is often a significant difference in the heat propagation rates between the solid and fluid phases. For instance, the solid phase, typically composed of metals with high thermal conductivity, can experience a much faster rate of heat transfer than the fluid phase, which is often air, water, or oil-substances with much lower thermal conductivity. This disparity in thermal conductivity between the phases can lead to the development of temperature gradients across the phases, causing localized regions where the solid matrix and fluid phase are not in thermal equilibrium. As a result, the LTE model becomes increasingly invalid in such situations, especially when transient effects are significant, such as during heating, cooling, or rapid phase change processes. In these cases, a more comprehensive approach, such as the local thermal nonequilibrium model, may be required to accurately account for the temperature differences between the solid and fluid phases. Hence, two energy equations for fluid phase and solid phase are given [25],

$$\varepsilon(\rho C_p)_f\left(\frac{\partial T_f}{\partial t}+\vec{V}.\,\vec{\nabla}T_f\right)=\varepsilon k_f\vec{\nabla}^2 T_f+h_{sf}a_{sf}\left(T_s-T_f\right) \tag{16}$$

$$(1-\varepsilon)\left(\rho C_p\right)_s\left(\frac{\partial T_s}{\partial t}+\vec{V}.\ \vec{\nabla} T_s\right)=(1-\varepsilon)k_s\vec{\nabla}^2 T_s-h_{sf}a_{sf}\left(T_s-T_f\right) \tag{17}$$

where $h_{sf}$is heat transfer coefficient between fluid phase and porous media, and it can be defined as $h_v/a_v.a_v$ is the volumetric heat transfer area, which is written by ε/r_h,and $r_h$is hydraulic diameter of the fluid phase ($r_h = V_f/a_{sf}$). Furthermore, $V_f$and$a_{sf}$are the volume of fluid and the interfacial area between fluid phase and porous media.

5. Conclusions and outlook

This chapter introduced two primary modeling methods that are commonly used to analyze transport phenomena in metal foams: the Pore Scale Method and the Volume Averaged Method. PSM involves directly solving the Navier–Stokes and energy equations at the pore level, offering highly accurate predictions of local flow and temperature fields. By averaging the governing equations over a representative elementary volume, VAM simplifies the problem and yields macroscopic transport equations, thereby significantly reducing computational complexity and enhancing suitability for engineering-scale problems. For heat transfer modeling, two important models are considered: Local Thermal Equilibrium and Local Thermal Non-Equilibrium. Under the LTE model, the solid and fluid phases are assumed to have equal temperatures at each location, allowing the use of a single energy equation. However, in transient conditions or systems with large differences in thermal conductivity, the LTE model becomes invalid. In such cases, LTNE models, which treat the solid and fluid temperatures separately and incorporate interfacial heat transfer terms, provide more accurate descriptions of the thermal behavior. This chapter also presents the governing equations corresponding to each modeling method and model, offering a foundational framework for simulating heat and fluid flow in porous metal structures.

Generally, metal foams have emerged as highly promising materials for enhancing heat transfer due to their unique combination of high porosity, large surface area, and excellent thermal conductivity. These features make them suitable for a wide range of applications such as compact heat exchangers, thermal energy storage systems, electronics cooling, and phase change material enhancement. However, despite their advantages, the study and practical application of heat transfer in metal foams present several significant challenges that continue to limit their broader implementation in industry. One of the primary challenges is the complex geometry of metal foams, which significantly complicates the modeling and simulation of heat and fluid flow. Unlike simple porous media, metal foams exhibit highly irregular, interconnected pore structures that vary at both the micro and macro scales. This makes it difficult to accurately represent their morphology in computational models and to establish reliable correlations for thermal and flow properties such as permeability, inertial coefficient, and effective thermal conductivity. Although pore-scale simulations using CFD or lattice Boltzmann methods can offer detailed insights, they are computationally expensive and often impractical for large-scale systems.

Despite these challenges, the outlook for metal foam in heat transfer applications remains highly promising. Advances in additive manufacturing (3D printing) now enable the fabrication of custom-designed foam structures with controlled porosity and topology, improving both performance and predictability. Similarly, ongoing research into hybrid materials such as metal foams coated with nanostructures or embedded with phase change materials, which opens new opportunities for multifunctional thermal systems. Furthermore, the development of multi-scale modeling techniques, coupled with machine learning for property prediction and optimization, holds the potential to overcome current computational barriers. As experimental techniques continue to evolve, providing more accurate measurements of thermal properties at the microscale, and as more validated LTNE models become available, the integration of metal foams in practical thermal systems is expected to accelerate. In conclusion, while heat transfer in metal foams is a field filled with technical hurdles, it is also rich in innovation potential. Addressing current limitations through interdisciplinary research combining materials science, thermal engineering, and computational modeling will pave the way for the next generation of high-performance, compact, and energy-efficient thermal management systems.

Conflict of interest

The authors declare no conflict of interest.

Appendices and nomenclature

a_{sf}	interfacial heat transfer area, m^2
a_v	volumetric heat transfer area, m^2/m^3
VAM	volume averaged method
C_p	specific heat capacity, J/kg K
CFD	computational fluid dynamics
FVM	finite volume method
FEM	finite element method
h_{sf}	interfacial heat transfer coefficient, W/m^2K
h_v	volumetric heat transfer coefficient, W/m^3K
K	permeability, m^2
k	thermal conductivity, W/mK
LBM	lattice Boltzmann method
LTNE	local thermal non-equilibrium
LTE	local thermal equilibrium
PSM	pore scale method
PCMs	phase change materials
p	pressure, N/m^2
q	heat flux, $W \cdot m^{-2}$
REV	representative elementary volume
r_h	hydraulic diameter of the fluid phase, m
T	temperature, K
t	time, s
V	velocity vector, m/s
β	thermal expansion coefficient, 1/K
ε	porosity
μ	dynamic viscosity, Pa.s
ρ	density, kg/m^3

Subscripts and superscripts

eff	effective
f	fluid
ref	reference
s	solid

Author details

Chunyang Wang[1*] and Moghtada Mobedi[2]

1 Institute of Engineering Thermophysics, Chinese Academy of Sciences, Beijing, China

2 Mechanical Engineering Department, Faculty of Engineering, Shizuoka University, Hamamatsu-shi, Japan

*Address all correspondence to: wangchunyang@iet.cn

References

[1] Caket GA, Wang C, Nugroho AM, Celik H, Mobedi M. Recent studies on 3D lattice metal frame technique for enhancement of heat transfer: Discovering trends and reasons. Renewable and Sustainable Energy Reviews. 2022;**167**:112697. DOI: 10.1016/j.rser.2022.112697

[2] Mobedi M, Wang C. Structural classification of PCM heat exchangers. In: Mobedi M, Hooman K, Tao W-Q, editors. Solid-Liquid Thermal Energy Storage: Modeling and Applications. 1st ed. Boca Raton: CRC Press; 2022. pp. 217-242. DOI: 10.1201/9781003213260

[3] Wang C, Mobedi M, Yang X, Shen Y, Zhao H, Chen H, et al. A comparison study of heat dissipation module between the consolidated and unconsolidated porous structures for thermoelectric cooler. Applied Thermal Engineering. 2024;**247**:123085. DOI: 10.1016/j.applthermaleng.2024.123085

[4] Heyhat MM, Mousavi S, Siavashi M. Battery thermal management with thermal energy storage composites of PCM, metal foam, fin and nanoparticle. Journal of Energy Storage. 2020;**28**: 101235. DOI: 10.1016/j.est.2020.101235

[5] Buonomo B, Celik H, Ercole D, Manca O, Mobedi M. Numerical study on latent thermal energy storage systems with aluminum foam in local thermal equilibrium. Applied Thermal Engineering. 2019;**159**:113980. DOI: 10.1016/j.applthermaleng.2019.113980

[6] Wang C, Mobedi M, Kuwahara F, Nakayama A. A numerical study on acceleration of melting process under forced convection by using high thermal conductive porous media. Heat Transfer Engineering. 2020;**41**:723-738. DOI: 10.1080/01457632.2018.1564204

[7] Wang C, Mobedi M. A new formulation for nondimensionalization heat transfer of phase change in porous media: An example application to closed cell porous media. International Journal of Heat and Mass Transfer. 2020;**149**:119069. DOI: 10.1016/j.ijheatmasstransfer.2019.119069

[8] Wang C, Sugiura T, Mobedi M, Chen H. Comparison of heat transfer enhancement between open and closed cell porous metal structures for solid–liquid phase change. International Journal of Numerical Methods for Heat & Fluid Flow. 2023;**33**:1797-1817. DOI: 10.1108/HFF-09-2022-0504

[9] Mobedi M, Hooman K, Tao W-Q. Structural Classification of PCM Heat Exchanger. 1st ed. Boca Raton: CRC Press; 2022. DOI: 10.1201/9781003213260

[10] Li W-Q, He Y-L, Tao W-Q. Experimental and numerical studies on melting phase change heat transfer in open-cell metallic foams filled with paraffin. Applied Thermal Engineering. 2012;**37**:1-9. DOI: 10.1016/j.applthermaleng.2011.11.001

[11] Wang C, Mobedi M. A comprehensive pore scale and volume average study on solid/liquid phase change in a porous medium. International Journal of Heat and Mass Transfer. 2020;**159**:120102. DOI: 10.1016/j.ijheatmasstransfer.2020.120102

[12] Wang C, Zheng X, Zhang T, Chen H, Mobedi M. The effect of porosity and number of unit cell on applicability of

volume average approach in closed cell porous media. International Journal of Numerical Methods for Heat & Fluid Flow. 2022;**32**:2778-2798. DOI: 10.1108/HFF-08-2021-0527

[13] Wang C, Mobedi M, Kuwahara F. Simulation of heat transfer in a closed-cell porous media under local thermal non-equilibrium condition. International Journal of Numerical Methods for Heat & Fluid Flow. 2019;**8**:2478-2500. DOI: 10.1108/HFF-01-2019-0081

[14] Wang C, Mobedi M, Kuwahara F. Analysis of local thermal non-equilibrium condition for unsteady heat transfer in porous media with closed cells: Sparrow number. International Journal of Mechanical Sciences. 2019; **157-158**:13-24. DOI: 10.1016/j.ijmecsci.2019.04.022

[15] Ozgumus T, Mobedi M, Ozkol U. Determination of Kozeny constant based on porosity and pore to throat size ratio in porous medium with rectangular rods. Engineering Applications of Computational Fluid Mechanics. 2014;**8**: 308-318. DOI: 10.1080/19942060.2014.11015516

[16] Celik H, Mobedi M, Nakayama A, Ozkol U. A study on numerical determination of permeability and inertia coefficient of aluminum foam using x-ray microtomography technique: Focus on inspection methods for reliability (permeability and inertia coefficient by tomography). Journal of Porous Media. 2019;**22**:511-529. DOI: 10.1615/JPorMedia.2019028887

[17] Zhao C, Opolot M, Liu M, Bruno F, Mancin S, Hooman K. Phase change behaviour study of PCM tanks partially filled with graphite foam. Applied Thermal Engineering. 2021;**196**:117313. DOI: 10.1016/j.applthermaleng.2021.117313

[18] Liu G, Xiao T, Guo J, Wei P, Yang X, Hooman K. Melting and solidification of phase change materials in metal foam filled thermal energy storage tank: Evaluation on gradient in pore structure. Applied Thermal Engineering. 2022;**212**: 118564. DOI: 10.1016/j.applthermaleng.2022.118564

[19] Pu L, Zhang S, Xu L, Ma Z, Wang X. Numerical study on the performance of shell-and-tube thermal energy storage using multiple PCMs and gradient copper foam. Renewable Energy. 2021; **174**:573-589. DOI: 10.1016/j.renene.2021.04.061

[20] Shakibi H, Afzal S, Shokri A, Sobhani B. Utilization of a phase change material with metal foam for the performance improvement of the photovoltaic cells. Journal of Energy Storage. 2022;**51**:104466. DOI: 10.1016/j.est.2022.104466

[21] Marri KG, Balaji C. Experimental and numerical investigations on the effect of porosity and PPI gradients of metal foams on the thermal performance of a composite phase change material heat sink. International Journal of Heat and Mass Transfer. 2021;**164**:120454. DOI: 10.1016/j.ijheatmasstransfer.2020.120454

[22] Briache A, Mahdaoui M, Boubaker R, Ouardouz M, Ahachad M, Harmand S. A comparative analysis of enhancement techniques for latent energy storage systems. International Journal of Heat and Mass Transfer. 2025;**248**:127224. DOI: 10.1016/j.ijheatmasstransfer.2025.127224

[23] Hu X, Gong X, Zhu F, Xing X, Li Z, Zhang X. Thermal analysis and optimization of metal foam PCM-based heat sink for thermal management of electronic devices. Renewable Energy. 2023;**212**:227-237. DOI: 10.1016/j.renene.2023.05.021

[24] Yang X, Wei P, Liu G, Bai Q, He Y-L. Performance evaluation on the gradient design of pore parameters for metal foam and pin fin-metal foam hybrid structure. Applied Thermal Engineering. 2020;**175**:115416. DOI: 10.1016/j.applthermaleng.2020.115416

[25] Nazari M, Mahmoudi Y, Hooman K. Introduction to fluid flow and heat transfer in porous media. In: Mahmoudi Y, Hooman K, Vafai K, editors. Convective Heat Transfer in Porous Media. 1st ed. Boca Raton: CRC Press; 2019. DOI: 10.1201/9780429020261

Chapter 2

CFD-Deep Learning Investigation of the Structure-Performance Relationship of Foam Reactor with Hierarchical Pore Structure for Dry Reforming of Methane

Shiqi You, Yixiong Lin, Ting Qiu and Chen Yang

Abstract

As a highly promising carbon capture and utilization approach, dry reforming of methane faces the critical challenge of carbon deposition arising from cold spots, thereby inducing catalyst deactivation. To address this, an innovative foam reactor with hierarchical pore structure, which has fine(small) and coarse(big) pores, was proposed to enhance the heat transfer process. Using numerical simulation and deep learning methods, this study explored how hierarchical pores affect heat transfer and DRM reactions. The simulation findings demonstrate that the overall heat transfer coefficient reaches an optimal value at a particular d_{coarse}/d_{fine} ratio when pumping power is kept constant. A synergistic heat and mass transfer mechanism was identified, and the optimal hierarchical structure boosted reaction performance by 14.1% (vs. uniform fine pores) and 13.0% (vs. uniform coarse pores).

Keywords: foam reactor, dry reforming of methane (DRM), hierarchical pore structure, deep learning, heat and mass transfer

1. Introduction

Greenhouse gas emissions give rise to abnormal global climate patterns. CO_2 and CH_4 are the primary GHGs, accounting for 73.5 and 21.6% of the total, respectively [1]. As an advanced catalytic process, methane dry reforming (DRM) represents a viable strategy to convert CO_2 and CH_4 into syngas (H_2/CO), with significant environmental and economic benefits [2–4]. However, the coke formation-induced catalyst deactivation is the main challenge restricting the widespread use of DRM [5].

DRM is highly endothermic; coke forms via CH_4 decomposition and Boudouard reaction at 557–700°C [5]. If the temperature is below 700°C, Boudouard reaction will occur in this process, whereas the optimal operational temperature range for DRM is from 870 to 1040°C [6]. Thus, high temperatures and enhanced heat transfer can enhance the reaction performance of DRM [7]. Consequently, upgrading heat transfer in DRM reactors is of crucial importance for mitigating the aforementioned challenges.

In practical DRM applications, fixed-bed and fluidized-bed reactors represent the most widely adopted reactor types across industrial-scale applications [8]. Fixed-bed reactors are extensively used in the chemical industry due to their structural simplicity. Yet, the random packing of catalyst particles within the reactor results in non-uniform bed void distribution, which in turn generates zones where heat transfer is constrained, especially along the radial axis. This not only reduces conversion of CO_2 and CH_4 but also sharply increases pressure drop while degrading heat transfer performance [9]. In contrast to fixed bed reactors, fluidized beds eliminate localized transfer limitations, resulting in enhanced thermal uniformity and mass transfer across the catalyst bed [10]. However, the accurate prediction of particle flow, heat transfer, mass transfer, and reaction kinetics is rendered highly problematic by the complex hydrodynamic phenomena and violent interparticle collisions inherent to fluidized-bed reactors, with this challenge serving as a pivotal barrier to the industrial scale-up of such reactor technology [11].

With a unique structural design comprising high porosity and a fully interconnected pore structure, the foam reactor stands out as a novel reactor type that can effectively disrupt the development of fluid boundary layers during reactive flow processes, significantly enhancing fluid mixing and radial mass and heat transfer [12, 13]. In a study of convective heat transfer with matched reactor sizes, Li et al. [14] observed that foam reactors exhibited more uniform radial and axial temperature distributions than fixed-bed reactors. Moreover, foam reactors have a high catalyst effectiveness factor [15]. Improving heat transfer and mass transfer in foam reactors requires optimization of pore size (PPI, pores per inch), porosity, and ligament shape, which are critical structure parameters for foam reactors [16–18]. For decades, this domain has stood as a thriving area of academic inquiry.

An innovative open-cell foam structure with a hierarchical pore structure, comprising fine (small) and coarse (big) pore sizes, was proposed by Lin et al. [19], aiming to enhance transport processes through structural optimization. In contrast to foam materials with a uniform pore structure, foams engineered with a hierarchical pore structure demonstrate markedly enhanced heat transfer performance. In a subsequent numerical investigation, Lin et al. [20] separately evaluated how the structural parameters of hierarchical pores influence the coking rate during the DRM reaction. Their numerical findings revealed that, during the DRM reaction, the rate of coke deposition in a foam reactor engineered with a hierarchical pore structure is diminished by 58% relative to that observed in the uniform pore structure.

Although numerical simulations afford the capability to conduct direct analyses of the phenomena associated with transport in porous media, numerical simulations often demand a lot of computing time and resources, particularly for large-scale domains. Therefore, developing an efficient alternative method is crucial for investigating transport processes. In recent decades, more and more researchers in chemical engineering are leveraging machine learning techniques to predict transport in the porous media [21, 22]. With respect to the deep neural network architectures employed for the predictive simulation of transport in porous media, the U-Net-based convolutional neural network (CNN) model is widely used for transport prediction in porous media [22–25]. Graczyk et al. [22] proposed the usage of the U-Net architecture CNN model, which accurately predicts the concentration fields in porous media. According to the research by Wang et al. [23], predictions of temperature fields and effective thermal conductivity for irregular porous media are achievable via a CNN model, with the model demonstrating high accuracy in capturing thermal transport behavior in complex porous structures. Notably, no existing studies have used CNN model to analyze transport in foam structures.

This research undertakes a comprehensive exploration of the structure-performance relationship exhibited by hierarchical pore structure of foam reactors during the DRM reaction. Capitalizing on the superior mass transfer characteristics of open-cell foam, we innovatively integrate deep learning with 3D numerical simulation to analyze fluid flow and heat transfer in open-cell foam featuring the hierarchical pore structures. These research findings establish a critical theoretical framework and provide actionable technical guidance for the optimized structural design and rational development of high-performance foam reactors tailored to industrial catalytic processes.

2. Mathematical model framework

2.1 General description

Within the porous domains of foam materials, fluid flow is a dominant transport process, while heat transfer falls into two primary categories: convective heat transfer in the interconnected pore network and conductive heat transfer through the foam's solid ligaments. Building on this fundamental understanding, the current study examines the integrated processes of transport and DRM reactions within a foam structure. As shown in **Figure 1**, DRM reaction takes place on the surface of solid foam ligaments, which serve the site for concurrent mass and heat transfer. Additionally, in terms of foam structure, our previous work proposed a new algorithm designed to generate foams featuring varied hierarchical pore structures. In this algorithm, d_{fine}, d_{coarse}, V_{coarse}, *and* V_{fine} denote the pore size of fine pore, the pore size of coarse pore, the volume of coarse pore, and the volume of fine pore, respectively.

2.2 Numerical scheme of MRT-LBM for foam reactor

The lattice Boltzmann method with multiple relaxation times (MRT-LBM), incorporating D3Q15 and D3Q7 discrete schemes, is employed to numerically simulate the fluid flow and heat transfer process in the foam reactor. The evolution equations are as follows [26, 27]:

$$f_\alpha(x + e_\alpha \Delta t, t + \Delta t) - f_\alpha(x,t) = -\left(M_f^{-1} S_f M_f\right)\left[f_\alpha(x,t) - f_\alpha^{eq}(x,t)\right] \tag{1}$$

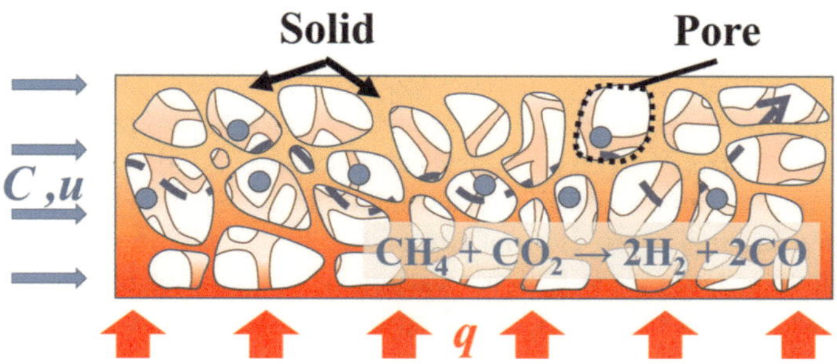

Figure 1.
Schematic illustration of transport and DRM reaction processes in a foam reactor.

Number	Reaction name	Reactions
1	Dry reforming of methane	$CO_2 + CH_4 \leftrightarrow 2CO + 2H_2$
2	Reverse water-gas shift	$CO_2 + H_2 \leftrightarrow CO + H_2O$
3	Coke formation	$CH_4 \leftrightarrow C + 2H_2$
4	Water-gas shift	$C + H_2O \leftrightarrow CO + H_2$
5	Boudouard reaction	$C + CO_2 \leftrightarrow 2CO$

Table 1.
Benguerba et al. [32] identified five reactions relevant to DRM [32].

$$g_a^i(x + e_a\Delta t, t + \Delta t) - g_a^i(x,t) = -\left(M^{-1}S_g^i M\right)\left[g_a^i(x,t) - g_a^{i(eq)}(x,t)\right] + w_a S_i \quad (2)$$

$$n_a(x + e_a\Delta t, t + \Delta t) - n_a(x,t) = -\left(M^{-1}S_h M\right)\left[n_a(x,t) - n_a^{eq}(x,t)\right] + w_a S_T \quad (3)$$

where t, x, and Δt are lattice time, spatial position, and time step. f_α is the density distribution, g_a^i is the mass concentration distribution, and n_a is the energy distribution. $f_\alpha^{eq}(x,t)$ is the equilibrium distribution function in e_α velocity. $M_f^{-1}S_f M_f$ is the collision matrix for D3Q15, in which M_f is transformation matrix, S_f is the diagonal matrix. $M^{-1}S_g^i M$ and $M^{-1}S_h M$ are the collision matrix for D3Q7, in which M is transformation matrix. S_g^i is a diagonal matrix. S_h is a diagonal matrix. The source terms (S_i, S_T) related to reaction, the reaction only occurs at this interface between the pore and solid skeleton. Therefore, S_i and S_T are zero in other areas.

2.3 Reaction kinetic data

Ni/Al_2O_3 catalyst exhibits good coke resistance in DRM reaction [28, 29]. In this work, the kinetics data of Ni/Al_2O_3 catalyst due to its extensive application in DRM reaction [30, 31]. Five reactions are considered on the Ni/Al_2O_3 catalyst, detailed descriptions of these reactions can be found in **Table 1**.

2.4 Convolutional neural network model

2.4.1 U-net architecture

Figure 2 presents the detailed architecture of the CNN model integrated with the U-Net structure, along with the computational workflow used to predict the temperature and velocity field across various foam configurations. The process starts with input parameters such as foam structure morphology and boundary conditions, ultimately generating output parameters that include detailed velocity and temperature fields. U-Net structure consists of two symmetric components: encoder and decoder layers. The encoder contains five identical convolutional blocks. Each network block incorporates three convolutional layers (kernel size of $3 \times 3 \times 3$), which are subsequently followed by the Rectified Linear Unit activation function to bolster the model's nonlinear characteristics. Following each individual block, a max pooling layer ($2 \times 2 \times 2$ kernel size) serves to extract salient features and reduce the dimensions of the input tensor. In a comparable manner, the decoder comprises five convolutional blocks, each mirroring the encoder's architectural design (three $3 \times 3 \times 3$ convolutional layers

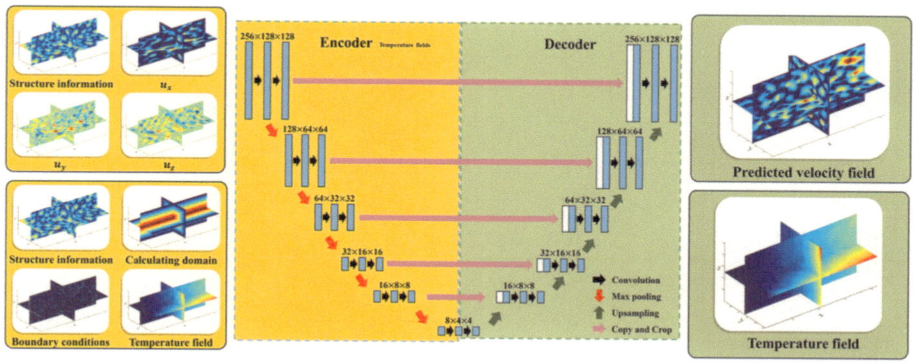

Figure 2.
Calculation process of CNN model.

coupled with an activation function). After every decoder block, a 2 × 2 × 2 upsampling layer—implemented through trilinear interpolation—undoes the compression of the encoder-derived features, thereby performing the decoding process. As shown in **Figure 2**, the input data are processed by the encoder, which progressively compresses it into lower-dimensional representations through a downsampling procedure. Subsequently, the decoder reconstructs the velocity and temperature fields from the latent representation by employing upsampling and convolution operations [33].

2.4.2 CNN model training and hyperparameter settings

In this study, the value of loss functions L_u for velocity distributions and L_T for temperature distributions are calculated as follows:

$$L_u = \frac{\sum \left| \left(u_i^{LBM} - u_i^{CNN} \right) \right|}{\sum \left| \left(u_i^{LBM} \right) \right|} \tag{4}$$

$$L_T = \frac{\sum \left(T^{LBM} - T^{CNN} \right)^2}{N_x N_y N_z} \tag{5}$$

where u_i^{LBM} and T^{LBM} denote the velocity in the i direction and temperature distributions computed using LBM, respectively, and u_i^{CNN} and T^{CNN} represent the corresponding velocity and temperature predictions obtained from CNN model. N_x, N_y, and N_z indicate the dimensions of the porous foam domain along the x, y, and z coordinate axes, respectively. To ensure model accuracy during training, the following configurations are implemented: a maximum epoch count of 1000 is defined; the batch size is fixed at 16 due to computational limitations; Initially set to 0.0001, the learning rate is scheduled to reduce by 50% per 200 epochs.

3. Verification and validation

3.1 Permeability in LBM

To evaluate the flow resistance of fluids in the foam reactor, permeability was determined using Darcy's law. In LBM simulation, a pressure gradient was applied

Sample number	PPI	ε	$K_D \times 10^9$ (m^2)		Deviation
			Present studies	Wu et al.	
1	10	0.76	21.2	23.3	8.8%
2	10	0.80	27.2	27.2	0.2%
3	10	0.84	34.9	32.7	6.7%
4	10	0.88	44.8	41.0	9.1%
5	10	0.92	60.4	55.1	9.7%
6	20	0.76	5.5	5.8	5.4%
7	20	0.80	7.0	6.8	3.2%
8	20	0.84	9.2	8.2	12.8%
9	20	0.88	11.4	10.3	9.5%
10	20	0.90	12.9	11.8	7.9%

Table 2.
Comparison of experimental and simulated permeability results.

along the x-direction by imposing density (pressure) differences between the inlet and outlet boundaries, with Zou-He scheme [34] employed to enforce these boundary conditions. The permeability was then calculated as follows:

$$K_D = \frac{Q\mu L}{A\Delta P} = \upsilon \bullet \frac{N_x \bullet \Delta x}{c_s^2 \bullet \Delta\rho} \bullet \frac{1}{N_x \bullet N_y \bullet N_z \bullet \Delta x^3} \sum_{i=1}^{N_x} \sum_{j=1}^{N_y} \sum_{k=1}^{N_z} u_x(i,j,k) \tag{6}$$

where K_D denotes the permeability of the foam structure, Q represents the flow rate, μ is the dynamic viscosity, L is defined as the characteristic length along the direction flow field, A refers to the cross-sectional area of the flow path, and ΔP signifies the pressure drop differential exerted across the foam structures. As shown in **Table 2**, the study compared two uniform pore structures (10 and 20 PPI) at different porosities, revealing strong consistency with Wu et al. [35]. The maximum observed deviation remained below 12.77%, validating the reliability of the findings.

3.2 Overall heat transfer coefficient in LBM

In this work, heat transfer occurs through two dominant mechanisms: convective heat transfer through the flow and conductive heat transfer through the solid foam ligaments. To quantitatively evaluate the thermal performance of the foam reactor, the overall heat transfer coefficient was determined using the following equation [36]:

$$h = \frac{\dot{q}}{A_{w,up}(T_{w,i} - T)} \tag{7}$$

$$\dot{q} = \dot{m}C_p(T_{out} - T_{inlet}) \tag{8}$$

$$T_{w,i} = \frac{T_{down} + T_{out}}{2} \tag{9}$$

where h is the overall heat transfer coefficient, $\dot{q}$ is the heat flux, $\dot{m}$ is the mass flux. $A_{w,up}$ is the bottom surface area for heating. C_p is the specific heat capacity. T_{out} *and* T_{inlet} are the average temperatures of the inlet and outlet fluid. $T_{w,i}$ is the

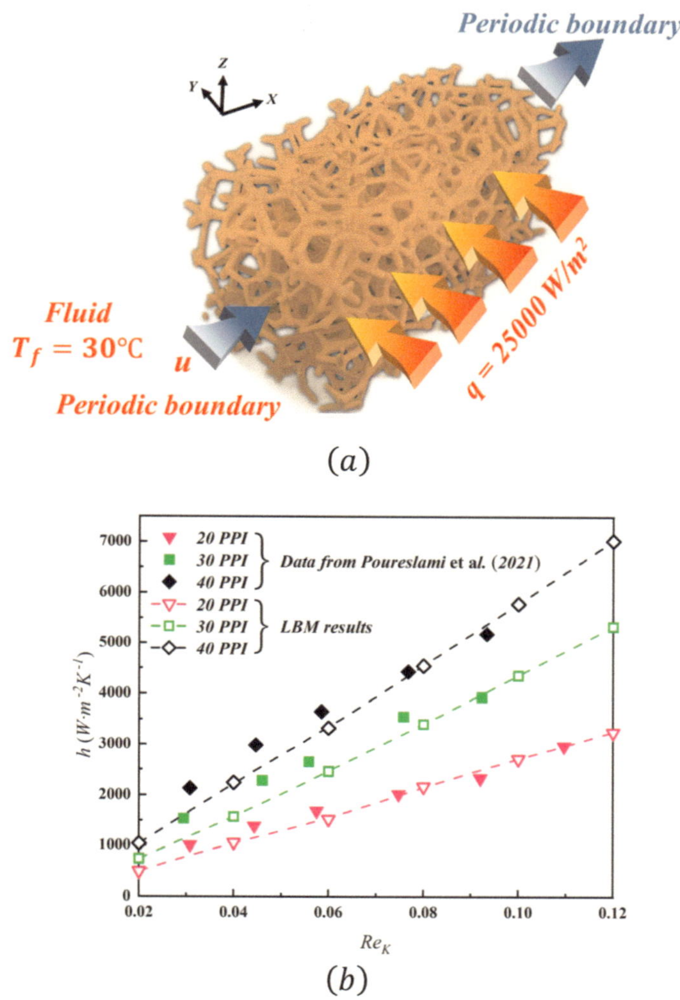

Figure 3.
(a) Boundary conditions for heat transfer process. (b) Comparison of h: Poureslami et al. [37] vs. this work.

mean surface temperature, which is the average temperature of the bottom and top surfaces. Systematic evaluation was performed on the overall heat transfer coefficients of three kinds of uniform foam structures with differing pore densities. The boundary conditions to calculate h, as well as a comparative analysis between the current results and Poureslami et al. [37] reports, are presented in **Figure 3(a)** and **(b)**, respectively. Excellent agreement between the two studies is demonstrated by the comparison, with the mean relative error calculated to stay below 15%.

4. Fluid flow and heat transfer characteristics in hierarchical pore structure

4.1 Fluid flow process

The CNN model was employed to investigate the fluid flow behavior within hierarchical pore structure. For CNN model training, 120 sets of lattice Boltzmann method (LBM) simulation data were utilized, incorporating the effects of three structural parameters on velocity distributions: the pore size ratio (d_{coarse}/d_{fine} = 1–3), the fine pore volume fraction (V_{fine}/V_{tol} = 0–100%), and the overall porosity (ε = 0.75–0.9). The pressure drop was fixed at $\Delta P = 2.1 \times 10^{-5}$ Pa during simulations. Additionally, the velocity direction distributions in the x-axis direction predicted by the CNN model are used to calculate the permeability, and the results were compared with those obtained from LBM simulations. As illustrated in **Figure 4**, the maximum absolute voxel-wise velocity errors in the two representative cases were approximately 4.1×10^{-5} m/s and 2.2×10^{-5} m/s. Furthermore, over half of the CNN model predicted velocities exhibited relative errors within 20%. These research findings demonstrate that the CNN model custom-developed in the present work delivers exceptional performance in predicting K_D of hierarchical pore foam. The CNN framework was also capable of effectively capturing the intrinsic structural characteristics of the foam matrix, a capability that forms the fundamental basis for its high-precision flow field prediction performance.

To expand the predictive capabilities of the model to key transport properties of porous media, the CNN model was further employed to forecast permeability across an array of hierarchical pore structures with distinct structural configurations. **Figure 5** investigates the influence of V_{fine}/V_{tol} and d_{coarse}/d_{fine} on permeability.

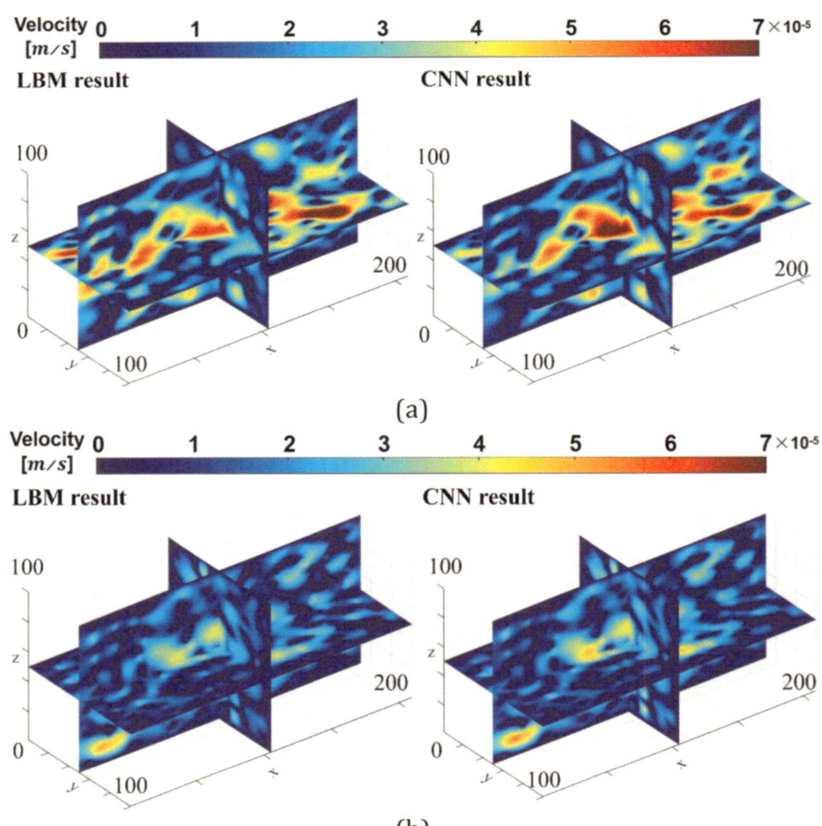

Figure 4.
Velocity distributions between CNN model and LBM model. (a) ε = 0.9, d_{coarse}/d_{fine} = 3.5, V_{fine}/V_{tol} = 50%; (b) ε = 0.9, d_{coarse}/d_{fine} = 3.5, V_{fine}/V_{tol} = 25%.

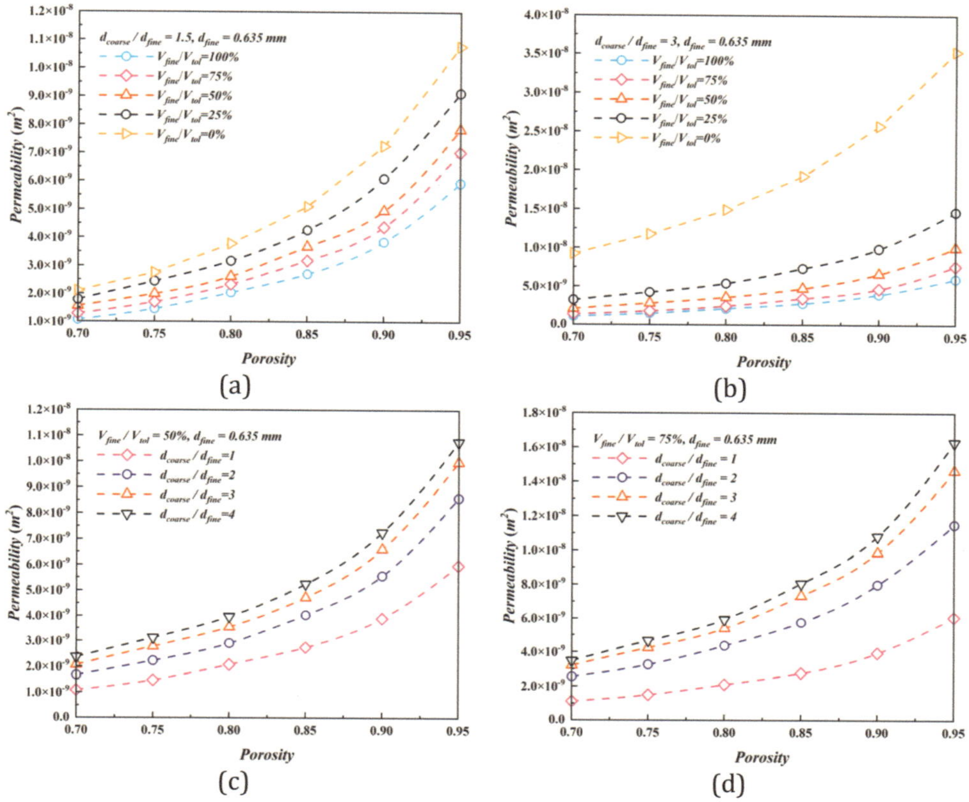

Figure 5.
(a, b): K_D predicted by the CNN model for different V_{fine}/V_{tol} ratio, (c, d): K_D predicted by the CNN model for different d_{coarse}/d_{fine} ratio.

As shown in **Figures 5a** and **b**, K_D increases drastically with the volume of coarse pores. This is consistent with the findings of Durmus et al. [38], who observed that hierarchical structures incorporating coarse pores into fine pores exhibit lower flow resistance compared to fine-pore foams at similar porosity levels, thereby validating our predictions. Regarding pore size ratio, **Figures 5c** and **d** show predictions for four ratios (1, 2, 3, 4) across porosities of 0.7–0.95. A clear upward trend emerges: under constant V_{fine}/V_{tol} and fixed d_{fine}, increasing coarse pore diameter enhances K_D.

4.2 Heat transfer process

Hundreds of sets of LBM simulation data were utilized for training the CNN model, which can predict temperature distributions. The training dataset incorporates three foam structural parameters. These parameters are pore size ratio (d_{coarse}/d_{fine} = 1–3.5), fine pore volume fraction (V_{fine}/V_{tol} = 0–100%), and overall porosity (ε = 0.85–0.95)—as well as the mean velocity (u_{mean} = 0.21–1.76 m/s). Furthermore, h was derived from the temperature distributions predicted by the CNN model, and the results were compared with those obtained from LBM simulations. The evaluation resulted in a mean relative error of 4.23% for h when comparing CNN predictions with LBM reference data, thereby confirming the model's predictive accuracy. **Figure 6** demonstrates strong agreement between LBM and CNN results.

The maximum absolute temperature errors are approximately 75.39 K (**Figure 6a**) and 74.86 K (**Figure 6b**), with the corresponding mean relative errors consistently below 10%.

The CNN model was further harnessed to generate predictions of h for different foam structures. **Figure 7a** shows predictions for five foam structures with V_{fine}/V_{tol} = 100, 67, 50, 25%. The results reveal two key trends: (1) the coefficient increases progressively with bulk mean velocity (u_{mean}). (2) Higher V_{fine}/V_{tol} ratios yield greater h. This occurs because finer pores increase specific surface area, whereas larger coarse pores reduce it – explaining the coefficient's decline as coarse pore volume grows. Pore size ratio (d_{coarse}/d_{fine}) also critically influences heat transfer. Predictions for four ratios (1, 1.5, 3, 4) in **Figure 7b** demonstrate that at fixed d_{fine} increasing coarse pore diameter (d_{coarse}/d_{fine}) reduces h. For practical evaluation under equal pumping power (pressure drop × flow rate [39]), **Figure 7c** reveals a critical distinction: while uniform pore structures (d_{coarse}/d_{fine} = 1) exhibit significant pressure drop increases with velocity, pressure drop in hierarchical structures especially show negligible changes (d_{coarse}/d_{fine}>2). As illustrated in **Figure 7d**, an optimal value of the heat transfer coefficient is observed when the coarse-to-fine pore size ratio (d_{coarse}/d_{fine}) falls within 2–3. Under the same h, the optimal hierarchical structure can reduce pumping power by approximately 30%.

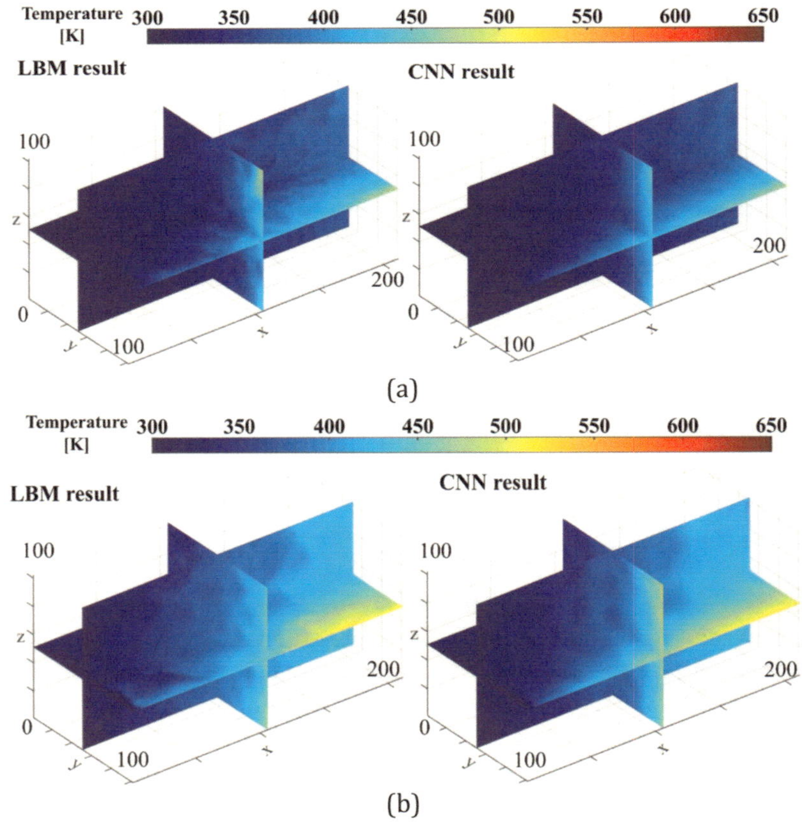

Figure 6.
Temperature distributions between LBM model and CNN model: (a) ε = 0.90, d_{coarse}/d_{fine} = 2, V_{fine}/V_{tol} = 50%, u = 2.01 m/s; (b) ε = 0.90, d_{coarse}/d_{fine} = 4, V_{fine}/V_{tol} = 50%, u = 1.01 m/s.

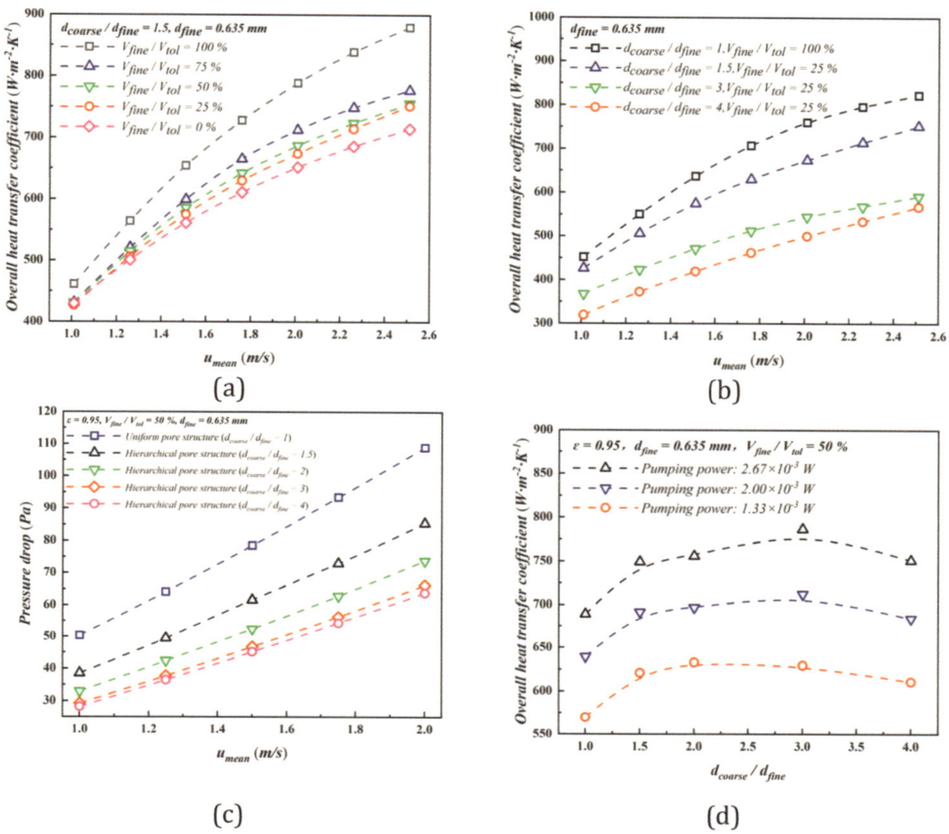

Figure 7.
(a) h predicted by the CNN model for different V_{fine}/V_{tol}; (b) h predicted by the CNN model for different d_{coarse}/d_{fine}; (c) ΔP with varying $d_{coarse}/$; (d) h predicted by CNN model of under different equal pumping power.

4.3 Computational time and robustness in CNN model

Table 3 shows the structure parameters domain for training data and testing data. The test sets are drawn from outside the training domain. It should be emphasized that the generation algorithm for foam structures possesses randomness. Even with the same structure parameters, foam structure may not be the same. The CNN model is employed to quantitatively predict the flow and temperature fields in distinct hierarchical pore foam structures that exhibit identical structural parameters. **Figure 8** shows the result of K_D and h, which also demonstrates that the CNN model has generalizability.

Computational efficiency is a critical performance metric for testing the CNN model. To study the advantage of the CNN methodology, we directly compare the computational time required for both training and testing phases against conventional Lattice Boltzmann Method (LBM) simulations. As detailed in **Table 4**, the prediction tasks encompass velocity and temperature distributions, utilizing 120 and 124 sets of training datasets, and 32 and 49 sets of test datasets, respectively. The results reveal a profound disparity in computational demands. For velocity distribution predictions, the CNN model achieves inference in a mere 0.132 minutes. This represents a significant contrast to the 17,280 minutes required by the LBM simulation, with the difference spanning several orders of magnitude. Remarkably, the efficiency gain is even more pronounced

Mode	Structure parameters	Train	Test
Velocity field mode	ε	0.75–0.95	0.7–0.95
	d_{coarse}/d_{fine}	1.0–3.0	1.0–4.5
	V_{fine}/V_{tol}	0–1	0–1
Temperature field mode	ε	0.85–0.95	0.7–0.95
	d_{coarse}/d_{fine}	1.0–3.5	1.0–4.0
	V_{fine}/V_{tol}	0–1	0–1
	u	0.21–1.77	0.21–2.01

Table 3.
The structure parameters domain for training data and testing data.

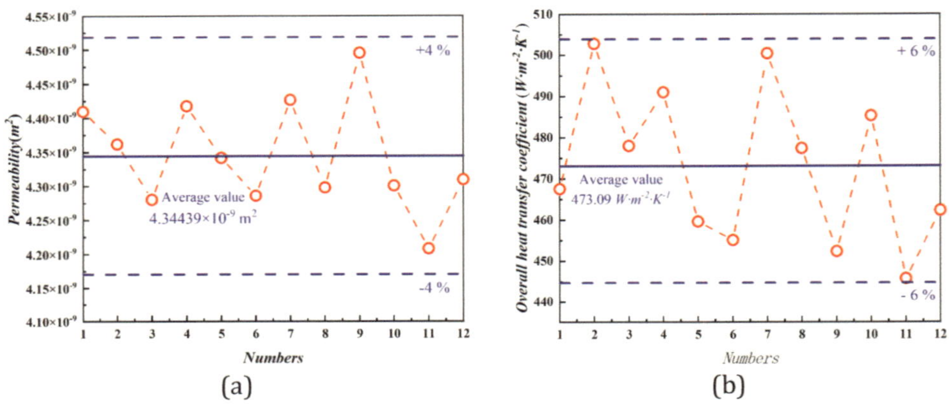

Figure 8.
K_D and h predicted by CNN model for a foam structure with hierarchical pore structure (ε = 0.85, d_{coarse}/d_{fine} = 3.5, V_{fine}/V_{tol} = 50%).

Calculation conditions		CNN training time	CNN prediction time	LBM simulation time
CPU (AMD 7B12)	Velocity distributions	—	—	17,280 min
	Temperature distributions	—	—	35,280 min
GPU (NVIDIA A800)	Velocity distributions	792 min	0.132 min	—
	Temperature distributions	150 min	0.252 min	—

Table 4.
The computational time consumed for foam structure simulations via the CNN and LBM model.

for temperature distributions. Training the CNN model on all 124 datasets requires only 150 minutes. Subsequently, performing predictions on new data consumes a negligible 0.252 minutes. Meanwhile, equivalent LBM simulations demand an extensive 35,280 minutes of computational effort. Moreover, as detailed in **Table 5**, the CNN model significantly reduces energy cost compared to LBM. Reducing computation from hundreds of hours to hours or minutes demonstrates the CNN model's superior suitability for rapid flow and thermal field predictions versus LBM.

Calculation conditions		CNN Training energy cost	CNN Prediction energy cost	LBM Simulation energy cost
CPU (AMD 7B12)	Velocity distributions	—	—	18.7 kW·h
	Temperature distributions	—	—	38.2 kW·h
GPU (NVIDIA A800)	Velocity distributions	5.8 kW·h	8.9×10^{-3} kW·h	—
	Temperature distributions	1.0 kW·h	1.7×10^{-3} kW·h	—

Table 5.
The computational energy cost for foam structure simulations via the CNN and LBM model.

4.4 DRM reaction in foam reactor with different hierarchical pore structures

Subsequent to an in-depth investigation into the heat transfer behaviors exhibited by hierarchical pore structure of foams, the reaction performance of DRM within such foams was subsequently investigated. To simplify the analysis, this study focuses solely on the initial coke formation rate for fresh catalysts, disregarding the influence of specific coke types on DRM reaction. To evaluate the reaction performance of the DRM system, a computational analysis was performed to determine the difference between the volume-averaged methane dry reforming reaction rate (R_{DRM}) and the volume-averaged coke formation rate (R_{Coking}). Specifically, R_{DRM} and R_{Coking} are defined and computed as follows:

$$R_{DRM} = \frac{\int_0^{S_{active}} r_1 dS}{V} \quad (10)$$

$$R_{Coking} = \frac{\int_0^{S_{active}} r_{Coking} dS}{V} \quad (11)$$

S and V represent the foam structure's specific surface area and computational domain volume, respectively. For this subsection, d_{fine} and porosity were held constant. A comprehensive study was carried out to explore the impact of hierarchical pore structures on the coke formation kinetics during the DRM reaction.

4.4.1 *Effect of V_{fine}/V_{tol} on DRM reaction*

For the purpose of specifically and methodically gauging the influence of hierarchical pore structures on the DRM reaction process, the effect of V_{fine}/V_{tol} in the foam reactor was investigated. With ε (0.95) and u y (1.51 m/s) kept constant, different V_{fine}/V_{tol} were tested. Meanwhile, the coarse-to-fine pore size ratio (d_{coarse}/d_{fine}) was fixed at 2. As shown in **Figure 9**, decreasing the V_{fine}/V_{tol} increases h in the foam structure, thereby enhancing heat transfer. Conversely, our previous work demonstrates that V_{fine}/V_{tol} reducing weakens mass transfer performance [20]. Therefore, controlling V_{fine}/V_{tol} achieves synergistic transport enhancement, yielding optimal reaction performance at V_{fine}/V_{tol} = 50%. **Figure 9** reveals two key trends: (1) The coking rate R_{Coking} consistently remained below the DRM reaction rate (R_{DRM}),

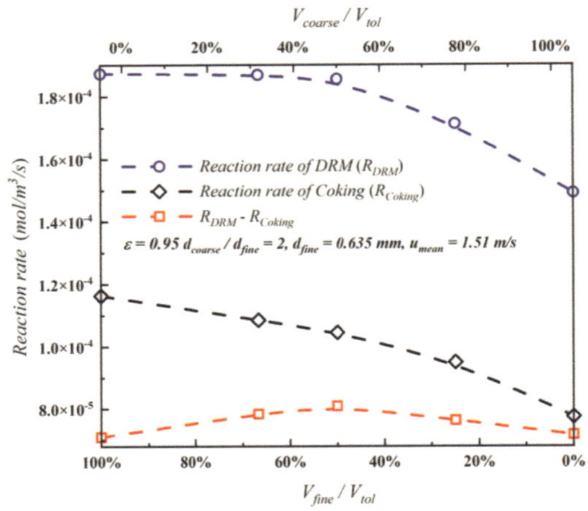

Figure 9.
The relationship between V_{fine}/V_{tol} and the apparent reaction rate of DRM. Adapted from Ref. [40].

exhibiting a clear decrease as V_{fine}/V_{tol} diminished; (2) R_{DRM} showed negligible variation above V_{fine}/V_{tol} = 50%, but declined markedly below this threshold. Consequently, the reaction performance demonstrates a non-monotonic trend, initially increasing and then decreasing with falling V_{fine}/V_{tol} peaking at V_{fine}/V_{tol} = 50%.

4.4.2 Effect of d_{coarse}/d_{fine} on DRM reaction

Based on the aforementioned findings, adjusting the hierarchical pore volume ratio yields an optimal foam structure with enhanced DRM reaction performance. Therefore, this section focuses on investigating the influence of pore size ratios (d_{coarse}/d_{fine}) on reaction performance, with the V_{fine}/V_{tol} fixed at 50%. Specifically, five distinct (d_{coarse}/d_{fine}) values were tested: 1, 1.5, 2, 3, and 4. As illustrated in **Figure 10**, both the DRM reaction rate (R_{DRM}) and coke formation rate (R_{Coking}) exhibit a gradual decreasing trend with an increase in d_{coarse}/d_{fine}. This trend can be attributed to the

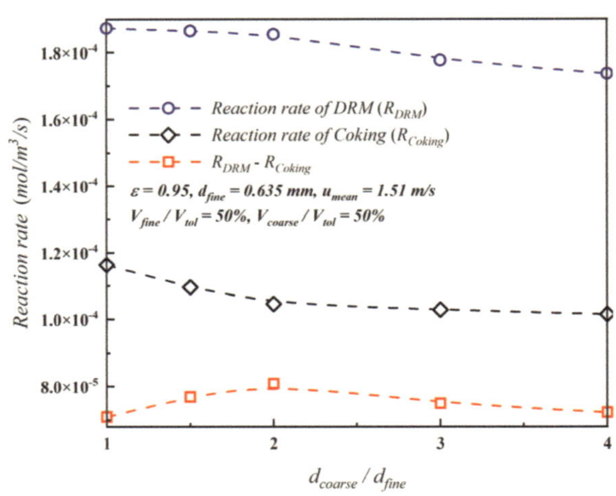

Figure 10.
The relationship between hierarchical pore diameter ratio and the apparent reaction rate of DRM. Adapted from Ref. [40].

reduction in the specific surface area of the foam structure, which weakens the catalytically active sites available for reactions. Furthermore, when d_{coarse}/d_{fine} decrease, the rate of decrease in R_{Coking} is higher than that in R_{Coking}; this observation thus signifies that the foam reactor achieves optimal reaction performance when the d_{coarse}/d_{fine} is 2.

5. Conclusions

In this study, the CNN model was trained to predict velocity and temperature distributions in hierarchical pore structure foam. The LBM was employed to generate training and testing datasets, and the flow and temperature fields predicted by CNN model can precisely calculate the permeability and overall heat transfer coefficient. Furthermore, this work systematically investigates the influence of hierarchical pore structures on transport and DRM reactions in foam reactors, with a specific focus on the effects of V_{fine}/V_{tol} and (d_{coarse}/d_{fine}) on coke formation rate. The findings demonstrate that the proposed CNN model can greatly expedite the design and scale-up of industrial foam reactors. The key results are summarized as follows:

1. The U-Net-based CNN model was developed for predicting velocity and temperature distributions in hierarchical pore foam. The model's validation results reveal mean relative errors of just 3.2% for permeability predictions and 4.2% for heat transfer coefficient predictions, highlighting its strong predictive capability across these two critical performance metrics. Notably, compared with LBM simulations, the CNN model reduces computational time by approximately 95.4% for velocity field prediction and 99.6% for temperature field prediction, demonstrating remarkable computational efficiency.

2. The trained CNN model was utilized to study the effects of V_{fine}/V_{tol} and d_{coarse}/d_{fine} on the permeability and heat transfer coefficient of hierarchical pore foam, respectively. Under constant pumping power conditions, the heat transfer coefficient reaches an optimal value when d_{coarse}/d_{fine} is in the range of 2–3.

3. By modulating V_{fine}/V_{tol} and d_{coarse}/d_{fine}, the optimal hierarchical pore structures were successfully delineated. This tailored structure demonstrates exceptional performance in the DRM reaction, leading to over 10% higher reaction rates compared to uniform pore structures.

Acknowledgements

This work was supported by the National Natural Science Foundation of China (nos. 22308058 and 52176062).

Author details

Shiqi You[1], Yixiong Lin[1,2*], Ting Qiu[1,2] and Chen Yang[1,2]

1 State Key Laboratory of Green and Efficient Development of Phosphorus Resources, Fujian Universities Engineering Research Center of Reactive Distillation Technology, College of Chemical Engineering, Fuzhou University, Fuzhou, Fujian, China

2 Qingyuan Innovation Laboratory, Quanzhou, China

*Address all correspondence to: yixionglin@fzu.edu.cn

References

[1] Filonchyk M, Peterson M, Zhang L, Hurynovich V, He Y. Greenhouse gases emissions and global climate change: Examining the influence of CO_2, CH_4, and N_2O. Science of the Total Environment. 2024;**935**:173359. DOI: 10.1016/j.scitotenv.2024.173359

[2] Budiman A, Song S, Chang T, Shin C, Choi M. Dry reforming of methane over cobalt catalysts: A literature review of catalyst development. Catalysis Surveys from Asia. 2012;**16**:183-197. DOI: 10.1007/s10563-012-9143-2

[3] Fabrik M, Salama A, Ibrahim H. Modeling of catalyst poisoning during hydrogen production via methane steam and dry reforming. Fuel. 2023;**347**: 128429. DOI: 10.1016/j.fuel.2023.128429

[4] Zhu Q, Zhou H, Wang L, Wang L, Wang C, Wang H, et al. Enhanced CO_2 utilization in dry reforming of methane achieved through nickel-mediated hydrogen spillover in zeolite crystals. Nature Catalysis. 2022;**5**:1030-1037. DOI: 10.1038/s41929-022-00870-8

[5] Jang W, Shim J, Kim H, Yoo S, Roh H. A review on dry reforming of methane in aspect of catalytic properties. Catalysis Today. 2019;**324**:15-26. DOI: 10.1016/j.cattod.2018.07.032

[6] Wang S, Lu G, Millar G. Carbon dioxide reforming of methane to produce synthesis gas over metal-supported catalysts: State of the art. Energy & Fuels. 1996;**10**:896-904. DOI: 10.1021/ef950227t

[7] Jang W, Jeong D, Shim J, Kim H, Roh H, Son I, et al. Combined steam and carbon dioxide reforming of methane and side reactions: Thermodynamic equilibrium analysis and experimental application. Applied Energy. 2016;**173**: 80-91. DOI: 10.1016/j.apenergy.2016.04.006

[8] Kaydouh M, El Hassan N, Chalhoub E, Bukharin N, El Hassan M. A review of computational fluid dynamics modeling of dry reforming of methane for improved catalyst and reactor design. Fuel. 2024;**366**:131290. DOI: 10.1016/j.fuel.2024.131290

[9] Teh L, Setiabudi H, Timmiati S, Aziz M, Annuar N, Ruslan N. Recent progress in ceria-based catalysts for the dry reforming of methane: A review. Chemical Engineering Science. 2021;**242**: 116606. DOI: 10.1016/j.ces.2021.116606

[10] Alotaibi F, Berrouk A, Salim I. Scaling up dry methane reforming: Integrating computational fluid dynamics and machine learning for enhanced hydrogen production in industrial-scale fluidized bed reactors. Fuel. 2024;**376**:132673. DOI: 10.1016/j.fuel.2024.132673

[11] Zambrano D, Soler J, Herguido J, Menéndez M. Conventional and improved fluidized bed reactors for dry reforming of methane: Mathematical models. Chemical Engineering Journal. 2020;**393**:124775. DOI: 10.1016/j.cej.2020.124775

[12] Yan P, Ding Q, Zou J, Na J, Geng X, Li H, et al. Numerical analysis of reaction kinetic accompanying flow and dispersion in structured ZSM-5@SiC open-cell foam catalyst. AIChE Journal. 2023;**69**:7. DOI: 10.1002/aic.18064

[13] Chen Y, Chen S, Shao Y, Quan C, Gao N, Fan X, et al. Siliceous mesocellular foam supported Cu catalysts for promoting non-thermal plasma activated CO_2 hydrogenation toward methanol synthesis. Frontiers of

Chemical Science and Engineering. 2024;**18**:77. DOI: 10.1007/s11705-024-2419-z

[14] Li Y, Zhang Q, Chai R, Zhao G, Liu Y, Lu Y. Structured Ni-CeO_2-Al_2O_3/Ni-foam catalyst with enhanced heat transfer for substitute natural gas production by syngas methanation. ChemCatChem. 2015;**7**:1427-1431. DOI: 10.1002/cctc.201500086

[15] Richardson J, Garrait M, Hung J. Carbon dioxide reforming with Rh and Pt-Re catalysts dispersed on ceramic foam supports. Applied Catalysis A-General. 2003;**255**:69-82. DOI: 10.1016/S0926-860X(03)00645-8

[16] Xiong J, Sun J, Chen Y, Nie Z. Study on the flow and heat transfer characteristics inside high-porosity open-cell copper foams: Experimental and numerical explorations. International Communications in Heat and Mass Transfer. 2024;**109**:109499. DOI: 10.1016/j.ijheatfluidflow.2024.109499

[17] Suo M, Sun K, Chen R, Che Z, Zeng Z, Li Q, et al. Oxygen transport in proton exchange membrane fuel cells with metal foam flow fields. Journal of Power Sources. 2022;**521**:230937. DOI: 10.1016/j.jpowsour.2021.230937

[18] Hu H, Zhao Y, Li Y. Research progress on flow and heat transfer characteristics of fluids in metal foams. Renewable and Sustainable Energy Reviews. 2023;**171**:113010. DOI: 10.1016/j.rser.2022.113010

[19] Lin Y, Yang C, Zhang W, Fukumoto K, Saito Y, Machida H, et al. Estimation of effective thermal conductivity in open-cell foam with hierarchical pore structure using lattice Boltzmann method. Applied Thermal Engineering. 2023;**218**:119314. DOI: 10.1016/j.applthermaleng.2022.119314

[20] Lin Y, Yang C, Zhang W, Machida H, Norinaga K. Lattice Boltzmann study on the effect of hierarchical pore structure on fluid flow and coke formation characteristics in open-cell foam for dry reforming of methane. Chemical Engineering Science. 2023;**268**: 118380. DOI: 10.1016/j.ces.2022.118380

[21] Yang G, Xu R, Tian Y, Guo S, Wu J, Chu X. Data-driven methods for flow and transport in porous media: A review. International Journal of Heat and Mass Transfer. 2024;**235**:126149. DOI: 10.1016/j.ijheatmasstransfer.2024.126149

[22] Graczyk K, Strzelczyk D, Matyka M. Deep learning for diffusion in porous media. Scientific Reports. 2023;**13**:9769. DOI: 10.1038/s41598-023-36466-w

[23] Wang M, Wang H, Yin Y, Rahardja S, Qu Z. Temperature field prediction for various porous media considering variable boundary conditions using deep learning method. International Communications in Heat and Mass Transfer. 2022;**132**:105916. DOI: 10.1016/j.icheatmasstransfer.2022.105916

[24] Wang H, Wang M, Yin Y, Qu Z. A universal structure of neural network for predicting heat, flow and mass transport in various three-dimensional porous media. International Journal of Heat and Mass Transfer. 2025;**241**:126688. DOI: 10.1016/j.ijheatmasstransfer.2025.126688

[25] Yu Y, Chen S, Wei H. Modified UNet with attention gate and dense skip connection for flow field information prediction with porous media. Flow Measurement and Instrumentation. 2023;**89**:102300. DOI: 10.1016/j.flowmeasinst.2022.102300

[26] Yoshida H, Nagaoka M. Multiple-relaxation-time lattice Boltzmann model for the convection and anisotropic

diffusion equation. Journal of Computational Physics. 2010;**229**:7774-7795. DOI: 10.1016/j.jcp.2010.06.037

[27] d'Humières D, Ginzburg I, Krafczyk M, Lallemand P, Luo L. Multiple-relaxation-time lattice Boltzmann models in three dimensions. Philosophical Transactions Series A Mathematical Physical and Engineering Sciences. 2002;**360**:437-451. DOI: 10.1098/rsta.2001.0955

[28] Huang Y, Li X, Zhang Q, Vinokurov VA, Huang W. Enhanced carbon tolerance of hydrotalcite-derived Ni-Ir/Mg(Al)O catalysts in dry reforming of methane under elevated pressures. Fuel Processing Technology. 2022;**237**: 107446. DOI: 10.1016/j.fuproc.2022.107446

[29] Gao X, Ge Z, Zhu G, Wang Z, Ashok J, Kawi S. Anti-coking and anti-sintering Ni/Al2O3 catalysts in the dry reforming of methane: Recent progress and prospects. Catalysts. 2021;**11**:1003. DOI: 10.3390/catal11081003

[30] Xie T, Zheng H, Xu K, Zhang Z, Yang B, Yu B. High performance Ni-based porous catalytically activated absorbers and establishment of kinetic model for complex solar methane dry reforming reaction system. Chemical Engineering Science. 2021;**239**:116625. DOI: 10.1016/j.ces.2021.116625

[31] Somaye S, Fereshteh M, Ali R, Mehran R. Influence of Fe, La, Zr, Ce, and Ca on the catalytic performance and coke formation in dry reforming of methane over Ni/MgO.Al_2O_3 catalyst. Chemical Engineering Science. 2022;**250**: 116956. DOI: 10.1016/j.ces.2021.116956

[32] Dehimi L, Benguerba Y, Virginie M, Hijazi H. Microkinetic modelling of methane dry reforming over Ni/Al_2O_3 catalyst. International Journal of Hydrogen Energy. 2017;**42**:18930-18940. DOI: 10.1016/j.ijhydene.2017.05.231

[33] Lin Y, Wu Z, You S, Yang C, Wang Q, Yin W, et al. Deep learning-based prediction of velocity and temperature distributions in metal foam with hierarchical pore structure. Green Chemical Engineering. 2025;**6**:209-222. DOI: 10.1016/j.gce.2024.08.003

[34] Zou Q, He X. On pressure and velocity boundary conditions for the lattice Boltzmann BGK model. Physics of Fluids. 1997;**9**:1591-1598. DOI: 10.1063/1.869307

[35] Wu Z, Caliot C, Bai F, Flamant G, Wang Z, Zhang J, et al. Experimental and numerical studies of the pressure drop in ceramic foams for volumetric solar receiver applications. Applied Energy. 2010;**87**:504-513. DOI: 10.1016/j.apenergy.2009.08.009

[36] Kurtbas I, Celik N. Experimental investigation of forced and mixed convection heat transfer in a foam-filled horizontal rectangular channel. International Journal of Heat and Mass Transfer. 2009;**52**:1313-1325. DOI: 10.1016/j.ijheatmasstransfer.2008.07.050

[37] Poureslami P, Siavashi M, Moghimi H, Hosseini M. Pore-scale convection-conduction heat transfer and fluid flow in open-cell metal foams: A three-dimensional multiple-relaxation time lattice Boltzmann (MRT-LBM) solution. International Communications in Heat and Mass Transfer. 2021;**126**:105465. DOI: 10.1016/j.icheatmasstransfer.2021.105465

[38] Durmus F, Maiorano L, Molina J. Open-cell aluminum foams with bimodal pore size distributions for emerging thermal management applications. International Journal of Heat and Mass Transfer. 2022;**191**:122852. DOI: 10.1016/j.ijheatmasstransfer.2022.122852

[39] Bai X, Zheng Z, Nakayama A. Heat transfer performance analysis on lattice core sandwich panel structures. International Journal of Heat and Mass Transfer. 2019;**143**:118525

[40] Lin Y, Yu M, Wang Q, Zhang W, Yin W, Yang C, et al. Unraveling the structure-performance relationship of foam reactor with hierarchical pore structure for dry reforming of methane reaction. Fuel. 2025;**389**:134596. DOI: 10.1016/j.fuel.2025.134596

Chapter 3

Enhancing Water Discharge and Output Performance in Proton Exchange Membrane Fuel Cell *Via* Metal Foam Flow Field with Hierarchical Pore Structure

Yun Sun, Yixiong Lin, Ting Qiu and Chen Yang

Abstract

To enhance the water discharge capability of metal foam flow fields possessing uniform pore structures in a proton exchange membrane fuel cell, a newly developed hierarchical pore structure is proposed for the metal foam flow field. This chapter numerically investigates the impacts of the structures of metal foam on water discharge capability, along with the output performance. Subsequently, 3D printing technology is utilized for fabricating metal foam flow fields with high precision. These fabricated flow fields are then experimentally compared with several other flow field configurations integrated into the cathode side. Experimental outcomes indicate that when applying a current density of 1.5 A/cm^2 over a duration of 3 hours, the volume of water discharged from the metal foam flow field that features a hierarchical pore structure approximates that of a parallel flow field. Specifically, this water discharge volume is 1.12 times and 1.30 times greater than those of the metal foam flow fields possessing uniform coarse pore and uniform fine pore, respectively. Furthermore, in comparison to the previously optimized approach, specifically the metal foam flow field possessing a pore per inch (PPI) of 75 and a compression ratio of 0.75, hierarchical pore structure can not only enhance the peak net power density by 9.5% and augment the water removal volume by 14.1%, but also reduce the pressure drop on the cathode side by two-thirds. This study establishes the theoretical foundation as well as provides technical perspectives for the application of metal foam flow fields in PEMFCs.

Keywords: PEMFC, metal foam flow field, hierarchical pore structure, water discharge capability, 3D printing

1. Introduction

Contemporary energy systems, heavily reliant on fossil fuels, have led to the depletion of natural reserves and adverse impacts on ecosystems [1], underscoring the urgent need to accelerate innovation in clean energy technologies [2]. Proton exchange membrane fuel cells (PEMFCs) have turned into a focal point within the realm of renewable energy research [3], with their exceptional performance efficiency, negligible environmental impact, and low acoustic emissions driving widespread adoption [4]. However, the broad-based commercial uptake of PEMFC is impeded by a multitude of hurdles [5]. To achieve successful market penetration, continuous improvements in performance, cost-effectiveness, and durability are essential [6].

In PEMFC, the bipolar plate (BP) is pivotal to its weight and cost, accounting for over 60% and 30%, respectively [7]. Optimizing the flow field design in bipolar plates substantially improves reactant distribution, water discharge capability, and fuel cell performance [8], representing a viable pathway to advance both technical performance and commercial feasibility of PEMFCs [9]. Nowadays, metal foam is gaining more attention as a flow field for PEMFCs [10], attributable to its superior reactant distribution characteristics, high thermal and electrical conductivity, as well as efficient heat removal [11].

Performance and mass transfer have been demonstrated to be improved by employing a metal foam flow field in PEMFC [12]. Ju et al. [13] performed a numerical investigation to examine the reactants transport efficiency and distribution properties of the metal foam and the classic serpentine flow channel. Their findings indicated that the metal foam configuration resulted in decreased reactant consumption as well as a more evenly distributed current density. Moreover, research findings by Awin and Dukhan [14] demonstrated that implementing metal foam flow fields in PEMFCs promotes better, even thermal dispersion throughout the membrane electrode assembly (MEA). Comparative analysis reveals that metal foam flow fields generate 9.9% higher maximum power density than classic serpentine designs. Though having immense real-world use potential in PEMFCs, according to Wan et al. [15], cathode-side application of metal foam flow fields results in a water management challenge at high current densities or elevated humidity levels. Improving water removal capability is essential for boosting the technology's practical implementation prospects.

Some researchers have applied hydrophobic materials to metal foam flow fields to improve water discharge capability [16, 17]. Baroutaji et al. [18] adopted polytetrafluoroethylene (PTFE) on the surface of metal foams to improve hydrophobicity and consequently enhance water removal performance. However, due to the limited conductivity of PTFE, this will generate a greater interfacial resistance at the interfacial zone between the flow field and the MEA. Moreover, several existing studies have systematically explored how structural characteristics affect liquid water removal dynamics within metal foam flow fields [19, 20]. According to Chen et al. [21], they adjusted the metal foam structure by means of compression, finding that a compression rate over 0.69 can boost water discharge and PEMFC performance. However, compression of metal foam flow fields results in increased parasitic power consumption. Through numerical simulation, Bao et al. [22] investigated pore size effects on water removal in metal foams, finding that drainage becomes challenging when droplet diameters approach pore dimensions. But the water removal efficiency can be enhanced when having much bigger or smaller pores than the droplets. The reason can be concluded that in larger pores, droplet separation from the metal foam's solid network can occur at a lower breakthrough velocity. While smaller pores make it easier to split water droplets. Therefore, designing a structure with more than two kinds of pore sizes can improve liquid water removal efficiency within metal foam flow fields. In their innovative work, Lin et al. [20] developed a newly designed hierarchical pore structure for metal foams, which incorporates both coarse pores and fine pores. Their results demonstrated a 17.6% enhancement in water removal efficiency relative to conventional uniform pore structure designs. Nevertheless, existing research has mainly investigated water management aspects, while the electrochemical characteristics of the hierarchical pore structure metal foam flow field remain unexplored within a complete PEMFC model.

This chapter explores the whole PEMFC performance by employing a newly designed metal foam flow field that features a hierarchical pore structure. Numerical simulations are employed to evaluate both the water removal process and power generation characteristics of the PEMFC. Following this, 3D printing is utilized to fabricate metal foam flow fields with high precision, enabling experimental comparison with conventional cathode flow-field designs in PEMFCs. Through systematic experimentation, the cell performance of metal foam flow field, parallel flow field, as well as compressed metal foam flow field configurations is rigorously evaluated [21].

2. Reconstruction of metal foam flow field

An innovative algorithm was applied to generate metal foam flow field structures that can yield both uniform and hierarchical pore structures. The algorithmic procedure is detailed below, with corresponding process steps visually presented in **Figure 1a**.

1. Construct a randomly packed assembly of two sphere classes with diameters d_{coarse} and d_{fine} using the discrete element method, where the particle counts are calculated according to the specified volume proportion V_{coarse}/V_{fine}.

2. Obtain the Cartesian coordinates for every sphere's geometric center. After removing the packed assembly, generate a 3D Voronoi structure derived from coordinate data.

3. Extract the nodal points and topological connections of all polyhedral cells in the Voronoi tessellation. Construct the solid matrix framework by interlinking these vertices following the elimination of the Voronoi structure.

4. Execute isotropic lattice expansion by progressively extending the solid matrix toward adjacent units according to the prescribed growth probability, iterating until reaching the specified porosity level (the porosity of all metal samples is 0.9).

Hierarchical metal foams are produced by this algorithm under non-unity diameter ratios ($d_{coarse}/d_{fine} \neq 1$). For enhanced simulation accuracy, the reconstruction algorithm incorporates necessary simplifications for mesh refinement. The spherical architecture is employed to substitute the polyhedral configurations generated through 3D Voronoi tessellation in step (2). This approach has been employed by Yu et al. [23] for metal foam fabrication. Subsequent applications include structural reconstruction by Sedeh et al. [24] for thermal conductivity analysis and by Xiao et al. [25] for fluid dynamics investigations in metal foams. Their numerically derived thermal conduction and fluid permeability parameters showed strong agreement with empirical results, demonstrating that the reliability of this simplified reconstruction

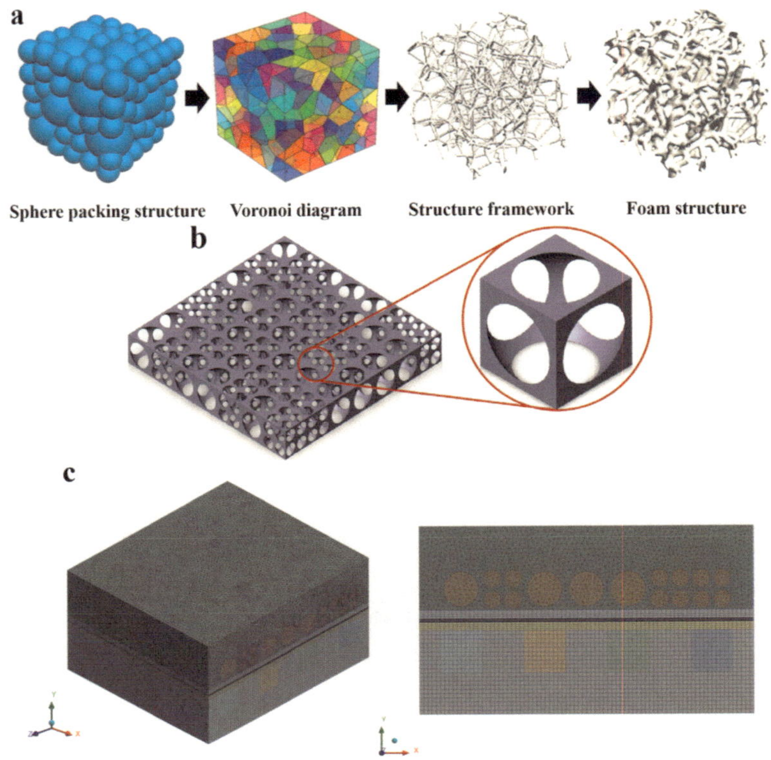

Figure 1.
(a) Reconstructed metal foam structure; (b) graphical presentation of the simplified foam configuration implemented in numerical modeling; (c) mesh representation of PEMFC possessing hierarchical pore structure.

approach works. Accordingly, the present study models metal foam pores as spherical structures for simplification. **Figure 1b** illustrates this conceptual representation of the metal foam architecture. The algorithm-designed metal foam configuration was applied to the cathode, contrasting with a conventional parallel-channel anode to create the PEMFC's geometric model. Moreover, the mesh view for PEMFC with a hierarchical pore structure model is also presented in **Figure 1c**.

3. Metal foam flow field within PEMFC

3.1 Numerical simulation

3.1.1 Physical models

This chapter investigates PEMFC performance through combined numerical and experimental analysis, focusing on water management, mass transfer, and electrochemical performance. Five distinct cathode flow field designs are explored. Specifically, they include metal foam flow field that features a uniform coarse pore design (case 1), one with a uniform fine pore design (case 2), another with a hierarchical pore structure design (case 3), traditional parallel flow field design (case 4), and compressed metal foam flow field design (case 5) [21] which can be observed in **Figure 2**. It is important to illustrate that the porosity of cases 1–3 is 0.9, and the porosity of case

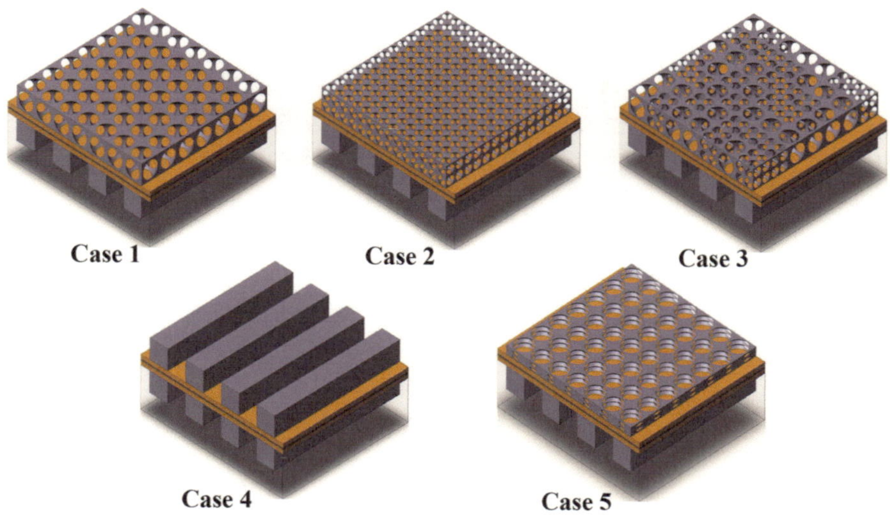

Figure 2.
Depiction of five distinct configurations of cathode flow fields: Case 1: Metal foam flow field that features a uniform coarse pore arrangement (d_{coarse} = 1 mm); Case 2: Metal foam flow field that features uniform fine pore arrangement (d_{fine} = 0.5 mm); Case 3: Metal foam flow field that features hierarchical pore arrangement (d_{coarse}/d_{fine} = 2; V_{coarse}/V_{fine} = 1; d_{fine} = 0.5 mm); Case 4: Parallel flow field (channel dimensions: 1 mm × 1 mm, rib width: 1 mm); Case 5: Compressed metal foam flow field (pore density: 75 PPI, compression ratio: 0.75).

5 after compression is 0.84. In the numerical simulation part, we compare three cases to study the mass transfer performance, water discharge processes, and cell performance within various metal foam flow fields. Afterward, during the experimental phase, these metal foam flow fields are fabricated with high precision *via* 3D printing technology. It is worth noting that the flow field structure and surface chemical properties of the metal foam will not change during the PEMFC reaction. In the subsequent phase, an assessment is made of the output performance of PEMFCs utilizing five distinct cathode flow fields.

3.1.2 Governing equations

To maintain computational efficiency while preserving predictive accuracy, the following assumptions were incorporated into our comprehensive three-dimensional, two-phase, isothermal fuel cell model:

1. A stable operational regime is maintained in the PEMFC.

2. The gaseous phase is treated as a noncompressible ideal fluid governed by laminar flow characteristics.

3. Isotropic behavior is assumed for all porous PEMFC elements (gas diffusion layer, catalytic layer, as well as membrane), with the membrane exhibiting zero gas permeability.

Table 1 summarizes the governing equations, while **Table 2** details their associated source items.

Description	Equations
Mass	$\nabla \cdot (\varepsilon \rho \boldsymbol{U}) = S_m$
Momentum	$\nabla \cdot (\varepsilon \rho \boldsymbol{U}\boldsymbol{U}) = -\varepsilon \nabla p + \nabla \cdot (\varepsilon \mu \nabla \boldsymbol{U}) + S_u$
Energy	$\nabla \cdot \left(\varepsilon s \rho_l c_{p,l} \boldsymbol{U} T + \varepsilon(1-s)\rho c_p \boldsymbol{U} T\right) = \nabla \cdot \left(k_{eff} \nabla T\right) + S_Q$
Species	$\nabla \cdot (\varepsilon \rho \boldsymbol{U} Y_i) = \nabla\left(\rho D_{i,eff} \nabla Y_i\right) + S_i$
	$D_{i,eff} = \varepsilon^{1.5}(1-s)^{r_s} D_{i,0}\left(\frac{p_0}{p}\right)^{\gamma_p}\left(\frac{T}{T_0}\right)^{\gamma_t}$
Charge	$\nabla \cdot \left(\sigma_{sol,mem} \nabla \phi_{sol,mem}\right) + R_{sol,mem} = 0$
	$R_{an} = \left(\xi_{an} j_{an}^{ref}(T)\right)\left(\frac{C_{H_2}}{C_{H_2}^{ref}}\right)^{\gamma_{an}}\left[e^{\alpha_{an} F \eta_{an}/RT} - e^{-\alpha_{ca} F \eta_{an}/RT}\right]$
	$R_{ca} = \left(\xi_{ca} j_{ca}^{ref}(T)\right)\left(\frac{C_{O_2}}{C_{O_2}^{ref}}\right)^{\gamma_{ca}}\left[-e^{\alpha_{an} F \eta_{ca}/RT} + e^{-\alpha_{ca} F \eta_{ca}/RT}\right]$
	$R_{sol} = \begin{cases} -R_{an} \text{ (anode)} \\ R_{ca} \text{ (cathode)} \end{cases}$
	$R_{mem} = \begin{cases} R_{an} \text{ (anode)} \\ -R_{ca} \text{ (cathode)} \end{cases}$
Dissolved phase	$\nabla \cdot \left(I_{ion} \frac{2.5}{22F} \lambda M_w\right) = \nabla \cdot (M_w D_\lambda \nabla \lambda) + S_w + S_{vl-d}$
	$D_\lambda = \begin{cases} 3.1\times10^{-7}\lambda\left(e^{0.28\lambda}-1\right)e^{(-2346/T)} & 0<\lambda\le 3 \\ 4.17\times10^{-8}\lambda\left(\lambda+161e^{-\lambda}\right)e^{(-2346/T)} & \lambda>3 \end{cases}$
Liquid water transport	$0 = \nabla \cdot \left(\frac{\rho_l K K_r}{\mu_l} \nabla (p_c + p)\right) + S_{v-l}$
	$p_c = \begin{cases} \frac{\sigma \cos\theta_c}{(K/\varepsilon)^{0.5}}\left[1.417(1-s) - 2.12(1-s)^2 + 1.263(1-s)^3\right] & \theta_c < 90^o \\ \frac{\sigma \cos\theta_c}{(K/\varepsilon)^{0.5}}\left[1.417s - 2.12s^2 + 1.263s^3\right] & \theta_c > 90^o \end{cases}$
	$\nabla \cdot (\rho_l \boldsymbol{U}_l s) = \nabla \cdot (D_l \nabla s)$

Table 1.
Governing equations applied in this work.

3.1.3 Boundary conditions

In this chapter, the PEMFC model's outer surfaces are designated with zero flux and act as immovable walls. In addition, the boundary conditions for the electrical potential are defined.

$$\begin{cases} \phi_{sol}=0 \text{ (terminal wall of the anode)} \\ \phi_{sol}=V_{cell} \text{ (terminal wall of the cathode)} \end{cases} \tag{1}$$

The flow field's inlet is set using mass flow rates, which are detailed below:

$$\bullet m_{inlet,an} = \frac{\rho_{an} \varsigma_{an} I_{ref} A_{MEA}}{2F} \frac{RT}{p_{an} - RH_{an} p_{sat}} \tag{2}$$

$$\bullet m_{inlet,ca} = \frac{\rho_{ca} \varsigma_{ca} I_{ref} A_{MEA}}{4F} \frac{RT}{0.21\left(p_{ca} - RH_{ca} p_{sat}\right)} \tag{3}$$

Source items	Expressions	Unit
Mass	$S_m = 0$ (Flow channel and GDLs)	kg/(m^3·s)
	$S_m = -\frac{M_{H_2}}{2F} R_{an}$ (Anode CL)	
	$S_m = \frac{M_{H_2O}}{2F} R_{ca} - \frac{M_{O_2}}{4F} R_{ca}$ (Cathode CL)	
Momentum	$S_u = -\frac{\mu}{K} U$ (GDLs and CLs)	kg/(m^2·s^2)
Energy	$S_Q = \frac{i_s^{\,2}}{\sigma_{sol}} - S_{v-l} \cdot L$ (GDLs)	W/m^3
	$S_Q = R_{an}\left(\eta_{an} - \frac{T\Delta S_{an}}{2F}\right) + \frac{i_s^{\,2}}{\sigma_{sol}} + \frac{i_m^{\,2}}{\sigma_{mem}} - (S_{v-l} + S_{vl-d}) \cdot L$ (Anode CL)	
	$S_Q = R_{ca}\left(-\eta_{ca} - \frac{T\Delta S_{ca}}{2F}\right) + \frac{i_s^{\,2}}{\sigma_{sol}} + \frac{i_m^{\,2}}{\sigma_{mem}} - (S_{v-l} + S_{vl-d}) \cdot L$ (Cathode CL)	
	$S_Q = \frac{i_m^{\,2}}{\sigma_{mem}}$ (PEM)	
	$S_Q = \frac{i_s^{\,2}}{\sigma_{sol}}$ (Current collector)	
Species	$S_{H_2} = -\frac{M_{H_2}}{2F} R_{an}$ (Anode CL)	kg/(m^3·s)
	$S_{O_2} = -\frac{M_{O_2}}{4F} R_{ca}$ (Cathode CL)	
	$S_{H_2O} = -\frac{M_{H_2O}}{2F} R_{ca}$ (CLs)	
Dissolved water	$S_w = \frac{M_w}{2F} R_{ca}$ (Cathode CL)	mol/(m^3·s)
	$S_w = 0$ (PEM, Anode CL)	
	$S_{vl-d} = \begin{cases} (1-s)\gamma_{vd} \frac{\rho_{mem}}{EW}(\lambda_{eq} - \lambda) \\ s\gamma_{ld} \frac{\rho_{mem}}{EW}(\lambda_{eq} - \lambda) \end{cases}$ (CLs)	
Liquid water	$S_{v-l} = \begin{cases} \gamma_e \varepsilon s \frac{M_w}{RT} p \ln\left(\frac{p - p_{sat}}{p - p_{wv}}\right) & p_{wv} \le p_{sat} \\ \gamma_e \varepsilon (1-s) \frac{M_w}{RT} p \ln\left(\frac{p - p_{sat}}{p - p_{wv}}\right) & p_{wv} > p_{sat} \end{cases}$	kg/(m^3·s)

Table 2.
Source items within governing equations.

where ρ_{an} and ρ_{ca} are considered to be the mixed gas densities for the anode and cathode. p_{an} as well as p_{ca} refer to the gas pressure entering the anode and cathode channels, respectively. Additionally, the key operating parameters along with physical properties employed in this work are compiled in **Table 3**.

3.1.4 Numerical methods and grid independence verification

Numerical solutions can be obtained by employing ANSYS Fluent's simulation environment. Numerical solutions are accepted only when the maximum residuals across all equations are attained magnitudes under 10^{-5}. Additionally, a grid independence is also performed as depicted in **Table 4**. Group 2 was chosen for further computations due to its optimal balance between computational cost and result accuracy, exhibiting less than 1% current density variation compared to Group 3.

3.1.5 Verification of the numerical model

To confirm the accuracy of the numerical model, the results of various flow field configurations are compared against the data in the existing study [30] as well as the

Parameter	Value	Reference
Open circuit voltage (V)	1.0	[26]
Mass fraction of H_2	0.112	[27]
Mass fractions of O_2 and H_2O	0.150/0.354	[27]
Operation temperature (K)	353.15	[27]
Stoichiometry ratio of anode/cathode	1.5/2.5	[27]
H_2 diffusivity (m^2/s)	1.1×10^{-4}	[27]
O_2 diffusivity (m^2/s)	3.2×10^{-5}	[27]
H_2O diffusivity (m^2/s)	7.35×10^{-5}	[27]
Other species diffusivity (m^2/s)	1.1×10^{-5}	[27]
Liquid water density (kg/m^3)	1000	[26]
Liquid coverage exponent	1	[27]
GDL/CL porosity	0.5/0.5	[27]
Contact angle of GDL, CL (°)	135	[27]
The permeability of GDL/CL	1×10^{-12}	[27]
Porosity of the metal foam flow fields	0.95	[20]
Anode/cathode reference exchange current density (A/m^2)	5000/0.05	[28]
Anode/cathode reference concentration (mol/m^3)	56.4/3.39	[28]
Anode/cathode concentration exponent	0.5/1.0	[29]
Equivalent weight of membrane (kg/kmol)	1.1×10^3	[29]
Faraday constant (C/mol)	96487	
Gas constant (J/(mol·K))	8.314	
Droplet size geometric factor	1×10^8	[30]
Water vapor/liquid water mass exchange rate constant	0.5/0.5	[30]

Table 3.
Operation conditions and key physical parameters.

Group no.	Grid number	Current density (A/cm^2)	Relative deviation (%)
Group 1	431252	1.44	2.78
Group 2	665277	1.48	-
Group 3	832674	1.49	0.68

Table 4.
Impact of grid number on the results.

current experimental outcomes when the operating parameters are the same. As evident in **Figure 3a**, the simulated polarization behavior using parallel flow fields closely matches reported numerical results. Next, regarding the PEMFC, a parallel flow field is equipped in the anode, while a metal foam flow field, which possesses a consistent pore size of 1 mm, is set in the cathode. We have also obtained and

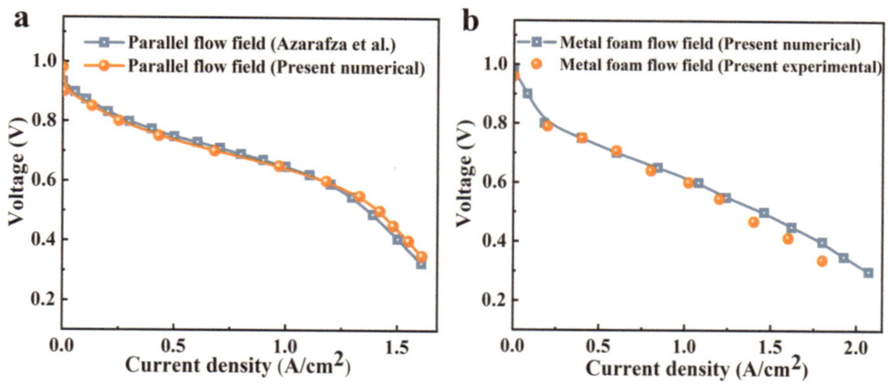

Figure 3.
(a) Comparison between current modeling results and Azarafza et al.'s simulations [30]; (b) comparison between current numerical and experimental results.

compared the current numerical results with the experimental data. **Figure 3b** shows a close match between experimental and simulation results at current densities below 1.2 A/cm^2. It is also worth noting that, given the current modeling constraints for capturing key experimental behaviors, including liquid water buildup and MEA performance decay. The computed values tend to be higher than the actual ones when the current density increases. Nevertheless, the maximum disparity in current density between the numerical and actual test outcomes amounts to merely 5.7%, and this value lies within an acceptable scope [31]. These findings demonstrate the reliability of the PEMFC model applied in this research.

3.1.6 Mass transfer and water discharge capability

The water management, along with the mass transfer features of cases 1–3, is detailedly investigated in this subsection. As depicted in **Figure 4a**, the mass-weighted mean velocity of case 2 is greater than that of cases 1 and 3 across the different voltage ranges. This effect is attributable to the mass transfer augmentation provided by small pore diameters of case 2. Furthermore, case 3 exhibits marginally reduced mass-weighted velocity compared to case 2, attributable to its localized fine pores. Still, fine pores are bad for water discharge as liquid water sticks more to fine pore ligaments. **Figure 4b** shows that at 1.8 A/cm^2, case 2 has the highest water molar concentration in the GDL-CL interface among cases 1–3, indicating poor water management. Case 3 (hierarchical pore structure) shows a marked reduction in water molar concentration compared to case 2. Our prior research shows that the coarse pore within the hierarchical pore structure is conducive to cutting the bond between liquid water and ligaments, thus improving water drainage. Also, because coarse and fine pores are randomly spread, the local gas velocity along the flow direction can be enhanced. So, case 3 has a better water discharge ability than cases 1 and 2, as shown in **Figure 4c**.

3.1.7 Output performance

After a systematic investigation of water discharge and mass transfer properties, this section further looks into the output performance of cases 1–3. **Figure 5a**

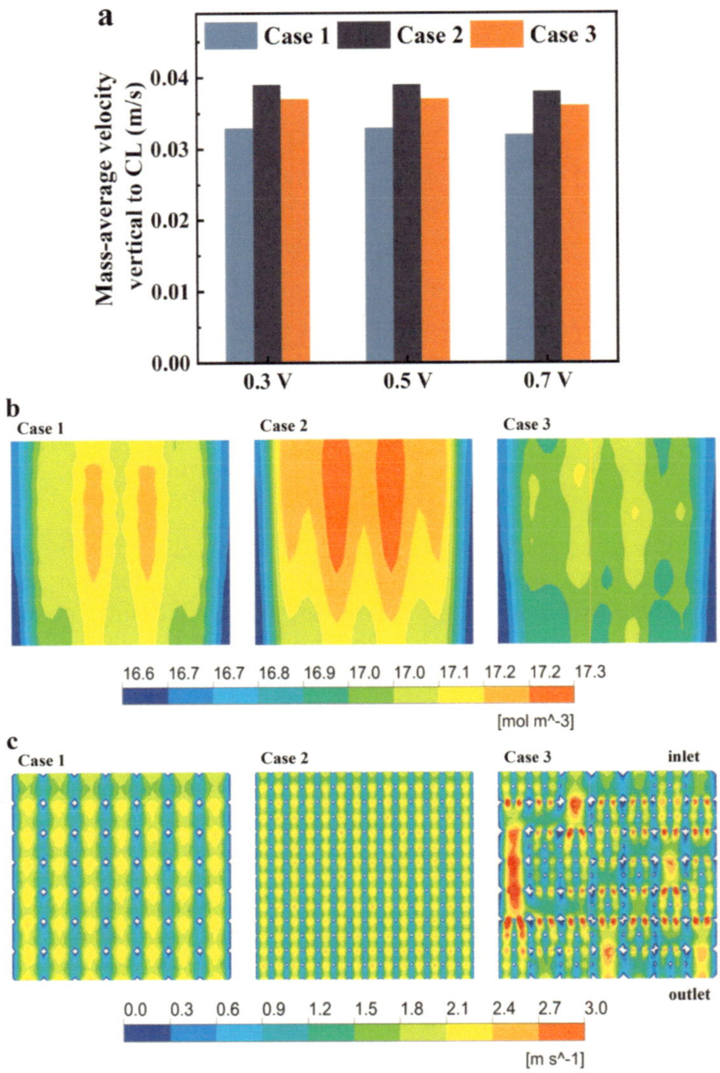

Figure 4.
(a) Mass-weight mean velocities perpendicular to the CL of cases 1–3 at the interface between GDL and flow field; (b) comparison of water molar concentration at the GDL-CL interface under 1.8 A/cm^2; (c) air velocity component along the direction from inlet to outlet under 1.8 A/cm^2.

demonstrates the I-V curves of the three cases. As can be observed, owing to the larger pore size, case 1 underperforms. Interestingly, case 3 featuring this novel hierarchical pore structure exhibits more excellent performance compared to case 2, especially under elevated current loads. For obtaining a clearer insight into the performance differences among cases 1–3, under 0.5 V, the oxygen concentration is also compared. From **Figure 5b**, it can be depicted that case 3 has the highest oxygen molar concentration distribution profile over the three cases, owing to its better mass transfer and water discharge performance. As clearly shown in **Figure 5c**, case 3 shows the more homogeneous current density, leading to enhanced power generation characteristics.

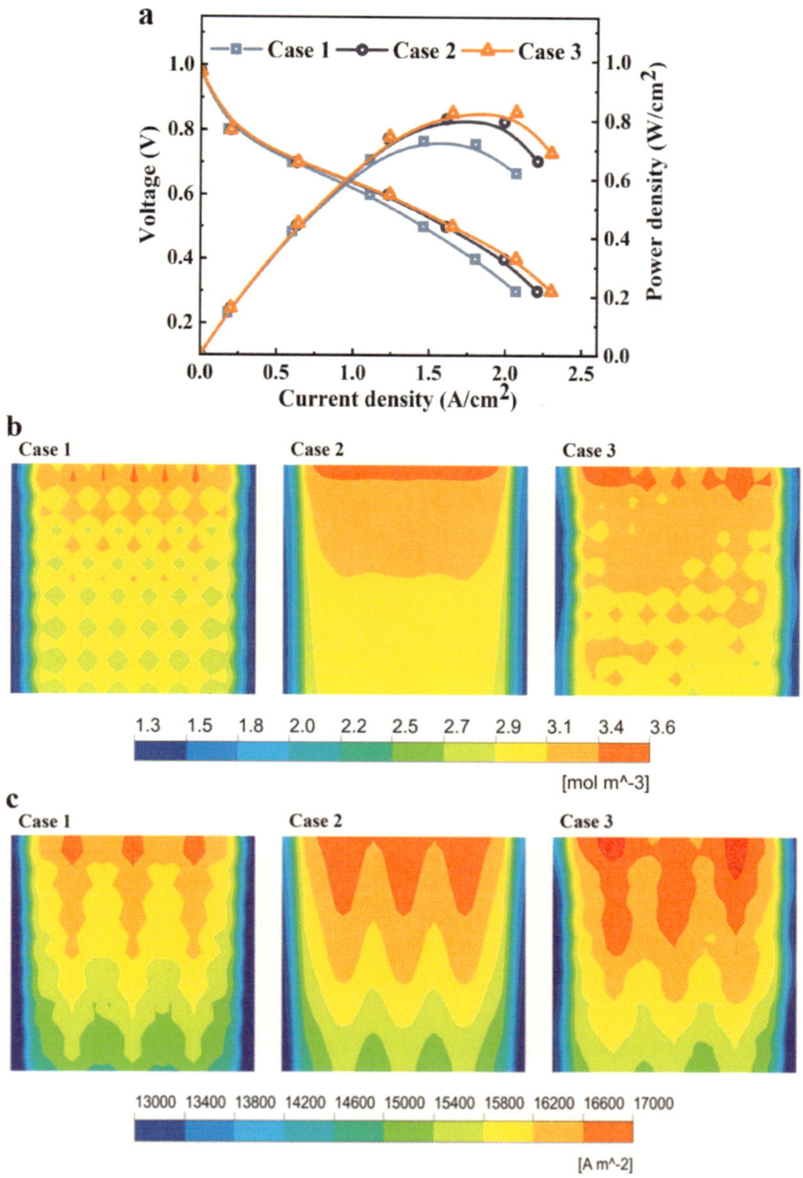

Figure 5.
(a) I-V curves of different cases; (b) Oxygen concentration distribution at the GDL-CL interface under 0.5 V; (c) Current density at the GDL-CL interface under 0.5 V.

3.2 Experiment

3.2.1 Fabrication of metal foam flow fields and PEMFCs integration

This chapter makes use of 3D printing technology to fabricate metal foam structures with high precision. **Figure 6a** illustrates the reconstruction process based on a computational algorithm developed in our earlier work [20]. Next, a metal 3D printer (iSLM160) is employed to construct the structure, using a nickel-base alloy (IN718) for metal foams. To guarantee the printed structure's accuracy, the microscopic

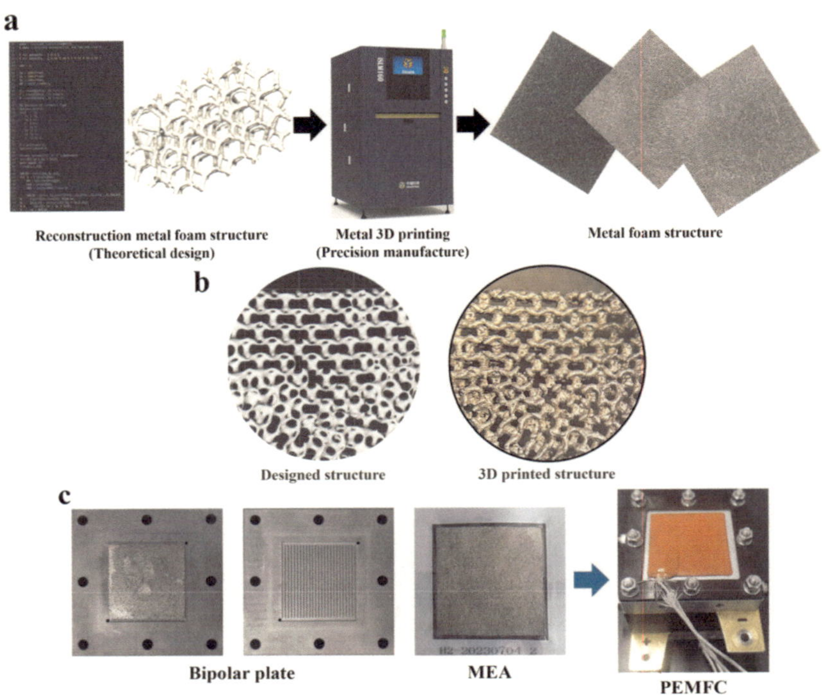

Figure 6.
(a) Schematic illustration of the process from designing to producing metal foam structures using a 3D printer; (b) comparison of structures between theoretical planning as well as actual fabrication; (c) PEMFC and its key parts.

structures of the metal foam are shown in **Figure 6b**. It is evident that the metal foam structures remain highly precise, covering both design and production stages. Following this, the PEMFC can be assembled. **Figure 6c** presents the completed PEMFC assembly alongside its key components, the bipolar plates and MEA.

3.2.2 Comparisons of experimental performance and water discharge

In this section, an experiment is carried out to evaluate the PEMFC performance and water removal across different flow-field designs. **Figure 7a** depicts that cases 1 and 4 perform worse than cases 2–3 due to their poor mass transfer. Case 3 delivers the top performance, particularly at elevated current density conditions. Looking at **Figure 7b**, it is evident that case 3 demonstrates the cathode pressure drop values intermediate to those of cases 1 and 2, aligning with the prior findings. Also, the water discharge amounts under 1.5 A/cm^2 over 3 hours are logged and contrasted to study different cases' water discharge abilities as depicted in **Figure 7c**. From **Figure 7d**, case 3's water discharge is 1.12 and 1.30 times higher than that of case 1 and 2. It is nearly equivalent to the conventional parallel flow field.

3.2.3 Comparison of different optimized strategies

To further verify the outstanding performance of the hierarchical pore structure, a comparison of the previous optimized strategy is carried out. The compressed metal foam flow field (75 PPI, being made from pure nickel and having a compression rate of 0.75) [21], named case 5, is taken into account in this section. From the comparison of the output performance presented in **Figure 8a**, it can be seen that

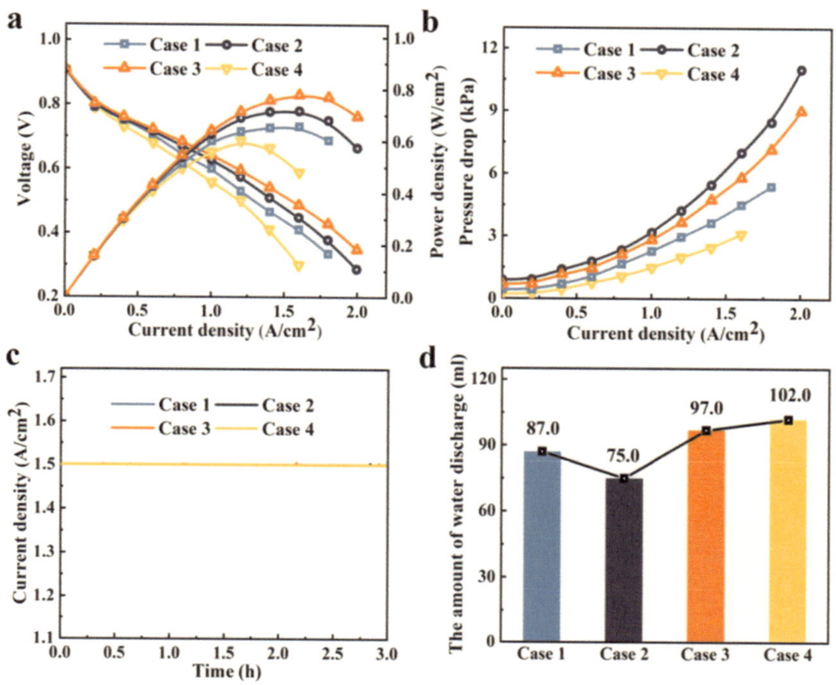

Figure 7.
Assessment of diverse flow field designs: (a) I–V curves; (b) pressure drop variation curves; (c) the amount of water discharge during 3 hours under 1.5 A/cm².

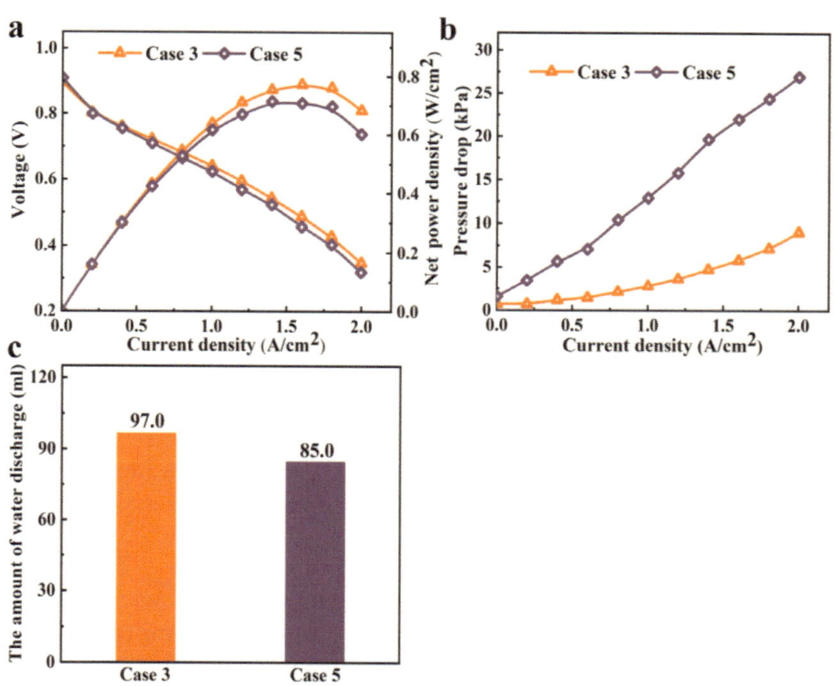

Figure 8.
Evaluation of distinct optimization approaches: (a) I-V curves; (b) pressure drop variation curves; (c) the amount of water discharge during 3 hours at 1.5 A/cm².

the peak net power density in case 3 is 9.5% greater compared to that in case 5. The definitions of the pumping power as well as net power density are given in Eqs. (4)–(6) [32]. Also, from **Figure 8b**, the cathode pressure drop in case 3, under 2 A/cm^{2,} amounts to just one-third of what it is in case 5. Moreover, it can be depicted from **Figure 8c** that the volume of water discharged in case 3 is around 14.1% greater compared to that in case 5 within 3 hours when applying a current density of 1.5 A/cm^2. This discovery suggested that this novel hierarchical pore structure demonstrates not only a reduced pressure drop but also excellent output performance.

$$P_{Net} = P_{cell} - P_{pump} \tag{4}$$

$$P_{cell} = I \bullet V \tag{5}$$

$$P_{pump} = \frac{u_{in} A_{in} \Delta p}{A_{MEA}} \tag{6}$$

Note that P_{Net} is the net power density, P_{cell} serves to indicate the cell power density, u_{in} is referred to as the inlet velocity, A_{in} denotes the region at the cathode inlet [33, 34].

4. Conclusions

In this chapter, a brand-new metal foam flow field that possesses a hierarchical pore structure was proposed for improving water discharge capability for PEMFCs. This research carried out a numerical exploration of the water discharge features and performance of metal foam flow fields with uniform pore structure and hierarchical pore structure. Moreover, 3D printing technology was utilized for accurately fabricating these metal foam flow fields. These fabricated metal foam flow fields were then contrasted with the parallel flow field and the compressed metal foam flow field [21]. After a comprehensive investigation, three conclusions were made:

1. The hierarchical pore structure demonstrates a bit worse mass transfer than in uniform fine pore ones. However, the water discharge in the hierarchical pore structure is nearly the same as that from the parallel flow field. Specifically, it is 1.12 times higher than in a uniform coarse pore structure and 1.30 times higher than in a uniform fine pore structure. This shows better water discharge efficiency in a hierarchical pore structure.

2. 3D printing is utilized for accurately fabricating the metal foam flow fields. Additionally, because of its outstanding mass transfer and water discharge features, a hierarchical pore structure can achieve superior output performance.

3. When contrasted with the prior optimization approach, namely the metal foam flow field characterized by 75 pores per inch and a compression ratio of 0.75, the metal foam flow field equipped with a hierarchical pore structure is capable of not only boosting the peak net power density by 9.5% and augmenting the water discharge by 14.1%, but also slashing the pressure drop by nearly two-thirds.

Acknowledgements

This work was financially supported by the National Natural Science Foundation of China (Nos. 22308058, 52176062).

Nomenclature

A_{MEA}	active area of membrane (m^2)
D	mass diffusivity (m^2/s)
I_{ref}	reference current density (A/m^2)
V	voltage (V)
K_r	relative permeability (m^2)
r_s	pore blockage exponent
L	characteristic length (m)
$j_{an,ca}$	exchange current density (A/m^3)
p_{sat}	saturated vapor pressure of water (Pa)
s	liquid water saturation
x	mass fraction (−)
Greek letters	
α	transfer coefficient (−)
η	local surface potential (V)
ε	porosity of porous area
ς	stoichiometry ratio (−)
θ_c	contact angle (°)
λ_{eq}	equilibrium water content
μ	dynamic viscosity (kg/(m·s))
σ	electricity conductivity (Ω/m)
ϑ	liquid coverage exponent
λ	dissolved water content

Author details

Yun Sun[1], Yixiong Lin[1,2*], Ting Qiu[1,2] and Chen Yang[1,2]

1 State Key Laboratory of Green and Efficient Development of Phosphorus Resources, Fujian Universities Engineering Research Center of Reactive Distillation Technology, College of Chemical Engineering, Fuzhou University, Fuzhou, Fujian, China

2 Qingyuan Innovation Laboratory, Quanzhou, China

*Address all correspondence to: yixionglin@fzu.edu.cn

References

[1] Gao B, Zhou Y. In-field experimental study and multivariable analysis on a PEMFC combined heat and power cogeneration for climate-adaptive buildings with taguchi method. Energy Conversion and Management. 2024; **301**(1):118003. DOI: 10.1016/j.enconman.2023.118003

[2] Asghar M, Lepikko S, Patakangas J, Halme J, Lund P. Comparative analysis of ceramic-carbonate nanocomposite fuel cells using composite GDC/NLC electrolyte with different perovskite structured cathode materials. Frontiers of Chemical Science and Engineering. 2018;**12**(1):162-173. DOI: 10.1007/s11705-017-1642-2

[3] Yang Y, Hao J, Ju C, Hao T, Wang X, Li Y, et al. Dynamic cross-scale modelling and thermos-electric response characteristics of PEMFC based on standardized thermal resistance. Energy Conversion and Management. 2024; **302**(15):118133. DOI: 10.1016/j.enconman.2024.118133

[4] Zhang H, Liang J, Xia B, Li Y, Du S. Ionic liquid modified Pt/C electrocatalysts for cathode application in proton exchange membrane fuel cells. Frontiers of Chemical Science and Engineering. 2019;**13**(4):695-701. DOI: 10.1007/s11705-019-1838-8

[5] Shi D, Cai L, Zhang C, Chen D, Pan Z, Kang Z, et al. Fabrication methods, structure design and durability analysis of advanced sealing materials in proton exchange membrane fuel cells. Chemical Engineering Journal. 2023;**454**:139995. DOI: 10.1016/j.cej.2022.139995

[6] Zhao D, Xia L, Dang H, Wu Z, Li H. Design and control of air supply system for PEMFC UAV based on dynamic decoupling strategy. Energy Conversion and Management. 2022;**253**:115159. DOI: 10.1016/j.enconman.2021.115159

[7] Yang C, Wan Z, Chen X, Kong X, Zhang J, Huang T, et al. Geometry optimization of a novel M-like flow field in a proton exchange membrane fuel cell. Energy Conversion and Management. 2021;**228**:113651. DOI: 10.1016/j.enconman.2020.113651

[8] Meng X, Ren H, Hao J, Shao Z. Design and experimental research of a novel droplet flow field in proton exchange membrane fuel cell. Chemical Engineering Journal. 2022;**450**:138276. DOI: 10.1016/j.cej.2022.138276

[9] Afshari E, Mosharaf-Dehkordi M, Rajabian H. An investigation of the PEM fuel cells performance with partially restricted cathode flow channels and metal foam as a flow distributor. Energy. 2017;**118**:705-715. DOI: 10.1016/j.energy.2016.10.101

[10] Park J, Lim J, Kim S, Choi I, Ahn C, Hwang W, et al. Enhancement of mass transport in fuel cells using three-dimensional graphene foam as flow field. Electrochimica Acta. 2018;**265**:488-496. DOI: 10.1016/j.electacta.2018.01.191

[11] Pourrahmani H, Moghimi M, Siavashi M. Thermal management in PEMFCs: The respective effects of porous media in the gas flow channel. International Journal of Hydrogen Energy. 2019;**44**:3121-3137. DOI: 10.1016/j.ijhydene.2018.11.222

[12] Liu R, Zhou W, Li S, Li F, Ling W. Performance improvement of proton exchange membrane fuel cells with compressed nickel foam as flow field structure. International Journal of Hydrogen Energy. 2020;**45**: 17833-17843. DOI: 10.1016/j.ijhydene.2020.04.238

[13] Jo A, Ju H. Numerical study on applicability of metal foam as flow distributor in polymer electrolyte fuel cells (PEFCs). International Journal of Hydrogen Energy. 2018;**43**:14012-14026. DOI: 10.1016/j.ijhydene.2018.01.003

[14] Awin Y, Dukhan N. Experimental performance assessment of metal-foam flow fields for proton exchange membrane fuel cells. Applied Energy. 2019;**252**:113458. DOI: 10.1016/j.apenergy.2019.113458

[15] Wan Z, Sun Y, Yang C, Kong X, Yan H, Chen X, et al. Experimental performance investigation on the arrangement of metal foam as flow distributors in proton exchange membrane fuel cell. Energy Conversion and Management. 2021;**231**:113846. DOI: 10.1016/j.enconman.2021.113846

[16] Tseng C, Tsai B, Liu Z, Cheng T, Chang W, Lo S. A PEM fuel cell with metal foam as flow distributor. Energy Conversion and Management. 2012;**62**: 14-21. DOI: 10.1016/j.enconman.2012.03.018

[17] Jang S, Park H, Jung J, Lee J, Park H, Jang J, et al. Enhanced water management of three-dimensional graphene-Ni foam with patterned wettability in a polymer electrolyte membrane fuel cell. ACS Sustainable Chemistry & Engineering. 2019;**7**: 15487-15494. DOI: 10.1021/acssuschemeng.9b03117

[18] Baroutaji A, Carton J, Stokes J, Olabi A. Application of open pore cellular foam for air breathing PEM fuel cell. International Journal of Hydrogen Energy. 2017;**42**:25630-25638. DOI: 10.1016/j.ijhydene.2017.05.114

[19] Bao Z, Niu Z, Jiao K. Numerical simulation for metal foam two-phase flow field of proton exchange membrane fuel cell. International Journal of Hydrogen Energy. 2019;**44**:6229-6244. DOI: 10.1016/j.ijhydene.2019.01.086

[20] Lin Y, Sun Y, Yang C, Zhang W, Wang Q, Wan Z, et al. Liquid water discharge capability enhancement of hierarchical pore structure in metal foam flow field of proton exchange membrane fuel cell. AIche Journal. 2024;**70**:e18256. DOI: 10.1002/aic.18256

[21] Chen X, Yang C, Sun Y, Liu Q, Wan Z, Kong X, et al. Water management and structure optimization study of nickel metal foam as flow distributors in proton exchange membrane fuel cell. Applied Energy. 2022;**309**:118448. DOI: 10.1016/j.apenergy.2021.118448

[22] Bao Z, Wang Y, Jiao K. Liquid droplet detachment and dispersion in metal foam flow field of polymer electrolyte membrane fuel cell. Journal of Power Sources. 2020;**480**:229150. DOI: 10.1016/j.jpowsour.2020.229150

[23] Yu Q, Thompson B, Straatman A. A unit cube-based model for heat transfer and fluid flow in porous carbon foam. Journal of Heat Transfer. 2006;**128**:353. DOI: 10.1115/1.2165203

[24] Sedeh M, Khodadadi J. Thermal conductivity improvement of phase change materials/graphite foam composites. Carbon. 2013;**60**:117-128. DOI: 10.1016/j.carbon.2013.04.004

[25] Xiao T, Yang X, Hooman K, Jin L, Yang C, Lu T. Conductivity and permeability of graphite foams: Analytical modelling and pore-scale simulation. International Journal of Thermal Sciences. 2022;**179**:107706. DOI: 10.1016/j.ijthermalsci.2022.107706

[26] Zhu K, Ding Q, Xu J, Yang Y, Yang C, Zhang J, et al. Optimization of

gas diffusion layer thickness for proton exchange membrane fuel cells under steady-state and load-varying conditions. Energy Conversion and Management. 2022;**267**:115915. DOI: 10.1016/j.enconman.2022.115915

[27] Shen J, Zeng L, Liu Z, Liu W. Performance investigation of PEMFC with rectangle blockage in gas channel based on field synergy principle. Heat and Mass Transfer. 2019;**55**(3):811-822. DOI: 10.1007/s00231-018-2473-5

[28] Cai G, Liang Y, Liu Z, Liu W. Design and optimization of bio-inspired wave-like channel for a PEM fuel cell applying genetic algorithm. Energy. 2020;**192**: 116670. DOI: 10.1016/j.energy.2019.116670

[29] Sun Y, Lin Y, Wang Q, Yang C, Yin W, Wan Z, et al. Novel design and numerical investigation of a windward bend flow field for proton exchange membrane fuel cell. Energy. 2024;**290**: 130142. DOI: 10.1016/j.energy.2023.130142

[30] Azarafza A, Ismail M, Rezakazemi M, Pourkashanian M. Comparative study of conventional and unconventional designs of cathode flow fields in PEM fuel cell. Renewable and Sustainable Energy Reviews. 2019;**116**: 109420. DOI: 10.1016/j.rser.2019.109420

[31] Zuo Q, Li Q, Chen W, Peng R, Zhu X, Xie Y, et al. Optimization of blocked flow field performance of proton exchange membrane fuel cell with auxiliary channels. International Journal of Hydrogen Energy. 2022;**47**: 39943-39960. DOI: 10.1016/j.ijhydene.2022.09.143

[32] Velisala V, Pullagura G, Chinnam N, Ganta R. Computational fluid dynamics study of a compound flow field for proton exchange membrane fuel cell (PEMFC) performance enhancement. Journal of Thermal Science. 2022;**31**(6): 2374-2384. DOI: 10.1007/s11630-022-1636-3

[33] Velisala V, Srinivasulu G. Numerical simulation and experimental comparison of single, double and triple serpentine flow channel configuration on performance of a PEM fuel cell. Arabian Journal for Science Engineering. 2018; **43**:1225-1234. DOI: 10.1007/s13369-017-2813-7

[34] Velisala V, Pullagura G, Golagani N. Numerical study of serpentine flow field designs effect on proton exchange membrane fuel cell (PEMFC) performance. Chemical Product and Process Modeling. 2021;**16**(1):55-66. DOI: 10.1515/cppm-2020-0023

Chapter 4

Foam Materials for Electrochemical Applications in Renewable Energy

Suchada Sirisomboonchai, Rongchi Dai and Koyo Norinaga

Abstract

This chapter provides a review of the design, fabrication, and electrochemical performance of foam-based materials as advanced electrodes for water splitting and electrochemical CO_2 reduction (CO_2RR). Foam structures ranging from metallic to carbon-based and hybrid composites offer high surface area, interconnected porosity, and excellent charge and mass transport properties. These attributes make them highly suitable for gas-evolving and multi-electron reactions such as the hydrogen evolution reaction (HER), oxygen evolution reaction (OER), and CO_2RR. This chapter highlights key fabrication strategies, including electrodeposition and templating, and discusses how foam morphology influences catalytic activity, gas release, and reactant accessibility. Emphasis is placed on structure-function relationships, particularly how pore architecture and catalyst distribution affect efficiency, selectivity, and stability. Additionally, this chapter concludes with current limitations and future prospects for implementing foam electrodes in scalable, real-world electrochemical systems.

Keywords: metal foam, electrochemical reaction, foam electrode, CO_2 reduction, water splitting

1. Introduction

The extensive utilization of fossil fuels has led to a persistent rise in global greenhouse gas emissions, thereby exacerbating climate warming and increasing the frequency of extreme weather events. To attain the objective of "carbon neutrality," current research priorities focus on advancing renewable energy technologies and optimizing their conversion into environmentally benign fuel vectors [1]. Electrochemical approaches offer significant advantages, including mild operating conditions, flexible integration with renewable energy sources (e.g., solar and wind), and highly controllable reaction kinetics. This facilitates the efficient management of multi-electron transfers and the regulation of intermediate species at ambient temperature and pressure, presenting a viable technological approach for clean energy conversion and carbon resource utilization [2]. For example, water electrolysis enables the direct conversion of renewable electricity into high-purity hydrogen while generating negligible greenhouse gas emissions under optimal operating conditions [3].

Similarly, electrocatalytic CO_2 reduction (CO_2RR) converts captured CO_2 into value-added chemicals (e.g., CO, formic acid, methanol, and ethylene) under ambient conditions. This process facilitates carbon recycling, mitigates environmental pressure from rising atmospheric CO_2, and yields significant economic returns [4]. However, achieving high performance and low cost in these electrochemical conversions depends on the development and refinement of advanced electrode materials that provide superior catalytic kinetics, reduced overpotentials, and enhanced operational durability [5].

Electrode materials are required to demonstrate high catalytic activity, superior electrical conductivity, adequate porosity, and mechanical stability. Although planar electrodes offer ease of fabrication, their flat topology results in uneven electrolyte transport and encourages the retention of gaseous products (e.g., H_2, O_2, and CO) on the surface. This retention leads to bubble formation, which creates diffusion barriers that increase the local overpotentials and obscure the active sites, thereby reducing the efficiency and selectivity. Conversely, three-dimensional (3D) porous electrodes, such as foam and hydrogel electrodes, utilize their inherently interconnected pore networks. This architecture allows for the distribution of numerous catalytically active sites throughout an expanded volumetric domain while simultaneously facilitating rapid electrolyte infiltration within the electrode matrix and promoting efficient bubble detachment and egress. Consequently, this significantly mitigates the adverse effects of bubble coverage and mass transport limitations, resulting in enhanced reaction kinetics and improved operational stability at high current densities [6].

Foam electrodes are three-dimensional porous structures composed of metal or carbon-based materials, fabricated through techniques such as chemical vapor deposition, electrochemical deposition, and mold casting. This porous "foam" structure typically exhibits a remarkably high specific surface area and a well-connected pore network, which facilitates the rapid diffusion of the electrolyte and provides numerous catalytically active sites. Foam electrodes can be categorized into two primary types based on the scaffold material: metal foams (e.g., copper and nickel foams) and carbon foams (e.g., carbon black and carbon nanotube-assembled foams). Metal foam electrodes demonstrate excellent electrical conductivity and can directly function as current carriers and catalytically active centers. Conversely, carbon-based foam electrodes can further support metal nanocatalysts, maintaining high conductivity, corrosion resistance, and relatively low cost [7].

Foam electrodes' 3D porous networks dramatically improve both hydrogen evolution and CO_2 reduction by preventing gas-bubble clogging and enhancing mass transport. In water electrolysis, rapid detachment of H_2 bubbles through internal pores avoids the resistive layer that elevates overpotential on planar surfaces, yielding higher current densities at lower voltages [8]. Similarly, in CO_2RR, interconnected pores facilitate CO_2 diffusion and adsorption deep within the foam while allowing swift expulsion of CO and C_2H_4 bubbles, thus reducing interface coverage and boosting current and Faradaic efficiency (FE) [9]. Moreover, alloyed metal foams (e.g., Ni-Co) or noble-metal-decorated foams further tune active-site electronics to lower overpotentials and enhance catalytic activity [6]. This unified foam-electrode design therefore delivers more efficient gas evolution, superior mass-transfer characteristics, and improved stability than conventional planar electrodes.

Foam electrodes, distinguished by their three-dimensional interconnected porous networks, offer substantial advantages for applications such as water splitting and CO_2 reduction (CO_2RR). Their architecture permits the adjustment of pore size and porosity through templating and deposition parameters, thereby maximizing the specific surface area while preserving mechanical strength. This structure facilitates efficient ion and gas transport, consequently minimizing both ohmic losses and concentration polarization [10]. Finally, the compatibility of foam-based systems with a variety of substrates and scalable fabrication techniques, such as templating, electrodeposition, and roll-to-roll (R_2R) processing, renders them highly promising for industrial-scale implementation [6].

2. Fundamentals of electrochemical reactions

Electrochemical catalytic technology has emerged as a crucial method for producing green hydrogen and valorizing carbon, primarily because of its ability to integrate effectively with renewable energy sources. This technology operates under ambient temperature and pressure conditions and offers superior process controllability. The hydrogen evolution reaction (HER) involves a multi-electron transfer process in which protons (H^+) or water molecules (H_2O) receive electrons at the electrode surface, resulting in the formation of molecular hydrogen (H_2). For example, in acidic electrolytes:

$$2H^+ + 2e^- \rightarrow H_2 \tag{1}$$

In an alkaline electrolyte:

$$2H_2O + 2e^- \rightarrow H_2 + 2OH^- \tag{2}$$

The catalyst is subject to stringent requirements concerning its electronic structure, surface adsorption energy, and electrical conductivity. The key performance indicators include the overpotential at 10 mA cm^{-2}, Tafel slope, and long-term stability under high current-density operation conditions [11]. The oxygen evolution reaction (OER), which constitutes the anodic half-reaction, involves a more complex mechanism characterized by four electron-proton-coupled steps and the formation/desorption of oxygenated intermediates (*OH, *O, and *OOH). This complexity imposes heightened demands on the oxidative resilience and surface-intermediate interactions of catalysts. In alkaline media, the OER can be represented as:

$$4OH^- \rightarrow O_2 + 2H_2O + 4e^- \tag{3}$$

In acidic electrolytes:

$$2H_2O \rightarrow O_2 + 4H^+ + 4e^- \tag{4}$$

Common performance metrics include the overpotential at 10 mA cm^{-2}, faradaic efficiency, and cycling stability [12]. By strategically modulating transition metal oxides, bimetallic layered double hydroxides (LDHs), and incorporating chiral molecules, researchers have effectively optimized the OER overpotential to a range of 160–220 mV [13].

The electrochemical CO_2 reduction reaction (CO_2RR) enables the transformation of captured CO_2 into a wide range of C_1–C_3 products, including carbon monoxide (CO), formic acid (HCOOH), methanol (CH_3OH), and ethylene (C_2H_4). This process involves a series of sequential steps, which include the adsorption and activation of CO_2, the formation of key intermediates such as *COOH and *OCHO, the cleavage of the C-O bond, and multi-electron coupling events [14]. The fundamental reaction pathways for the production of CO, formic acid, and C_2 + products are delineated as follows:

$$CO_2 + H^+ + 2e^- \rightarrow CO + H_2O \tag{5}$$

$$CO_2 + 2H^+ + 2e^- \rightarrow HCOOH \tag{6}$$

$$CO_2 + 6H^+ + 6e^- \rightarrow CH_3OH + H_2O \tag{7}$$

$$CO_2 + 8H^+ + 8e^- \rightarrow CH_4 + 2H_2O \tag{8}$$

$$2CO_2 + 12H^+ + 12e^- \rightarrow C_2H_4 + 4H_2O \tag{9}$$

$$2CO_2 + 12H^+ + 12e^- \rightarrow C_2H_5OH + 3H_2O \tag{10}$$

The choice of electrode material and its architecture significantly impacts the distribution of the local electric field, interfacial pH, and energy barriers associated with critical reaction intermediates. These factors ultimately determine the selectivity and distribution of reduction products [15]. The precise regulation of active sites and the design of electrode microenvironments, such as local proton concentration and electric field distribution, have emerged as key areas of research for enhancing the performance of CO_2RR [16].

3. Foam materials and fabrication method

Foam materials are a new class of materials with outstanding properties that make them suitable for a wide range of applications. These materials possess excellent mechanical properties and are lightweight, while still maintaining high strength and stiffness. In addition, foam materials exhibit superior acoustic properties, making them effective for sound absorption. The complex internal geometry of foam structures increases the surface area per unit volume, which is highly desirable for heat transfer or thermal management applications, such as heat exchangers. The characteristics of metal foams such as permeability, pore size, and pores per inch (PPI) can be tuned to suit specific applications [17, 18]. Since 2015, scientific publications related to foam-based electrodes have shown a steady increase. Although research on this topic was relatively limited before 2015, the number of publications has grown substantially each year. Notably, interest in their applications for renewable energy such as water splitting and electrochemical CO_2 reduction has continued to rise, reflecting the growing importance of sustainable energy technologies (**Figure 1**).

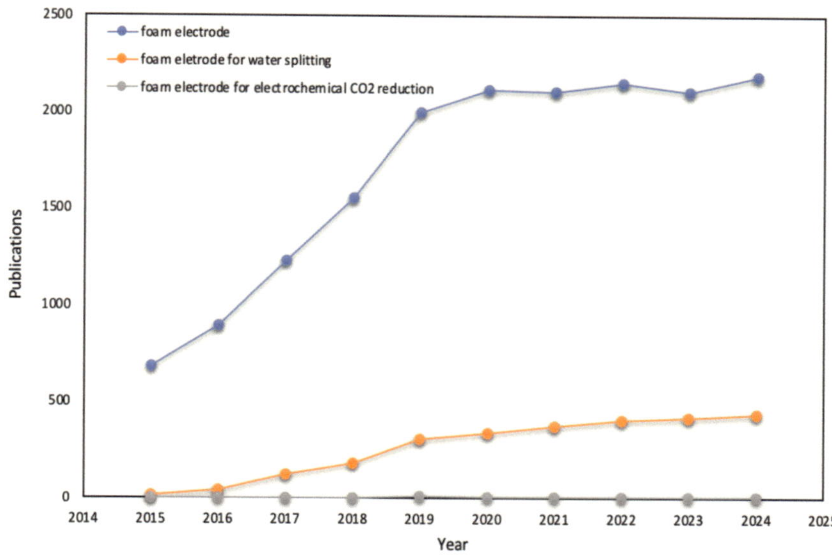

Figure 1.
Trends in the number of publications related to foam electrodes from 2015 to 2024, data retrieved from Web of Science using relevant keyword searches conducted in May 2025 (Search terms used included "foam electrode," "foam electrode for water splitting," and "foam electrode for electrochemical CO_2 reduction." Only articles indexed in the Web of Science Core Collection were considered.

Foam materials can generally be classified into two categories: *open-cell foams* and *closed-cell foams*. Open-cell metal foams allow fluids to pass freely from one cell to another, as the cells are interconnected and not sealed off. These foams are typically used in functional applications where load-bearing is not the primary concern [19]. In contrast, *closed-cell metal foams* consist of continuous cell walls that separate each cell from the others, typically forming spherical shapes [20]. These materials offer good mechanical properties but restrict access to the inner surfaces of the foam, making them more suitable for structural or load-bearing applications. The selection of foam with specific characteristics should be based on the design requirements of the intended application [21]. While closed-cell foams excel in mechanical performance, open-cell foams are generally necessary for applications involving internal interactions, such as fluid or gas flow. *Foam electrodes*, such as those made from metal or carbon, are widely used in electrocatalysis due to their highly porous, three-dimensional structure. Metal foams are often more suitable for use as electrodes than carbon black because their porous nature enhances catalytic activity and reduces transport resistance within the system.

3.1 Metal foam

3.1.1 Noble metals foam (NMF)

Noble metals foam (NMF), which include gold (Au), silver (Ag), ruthenium (Ru), palladium (Pd), iridium (Ir), and platinum (Pt), are known for its superior resistance to oxidation and corrosion. Self-supported macroscopic foams composed of noble metals have been synthesized using single or bimetallic systems, such as Au, Ag, Pt, Au-Ag, and Ag-Pt. These foams comprise 3D interconnected networks formed by metallic nanowires (NWs). NMFs can be engineered with various compositions and morphologies, including monometallic and multi-metallic alloys. The nanostructures may form nanospheres, nanowires, nanotubes, nanosheets, nanocubes, or core-shell nanoparticles. The nanowire diameter ranges from ~3 nm to over 1000 nm, with bulk densities from 1.3 to 773 mg·cm^{-3} and specific surface areas of 0.1 to 269 m^2·g^{-1} [22, 23]. However, NMFs face challenges including limitations in structural control, difficulties in obtaining high-purity products, and high precursor costs.

3.1.2 Nickel foam

Nickel foam is a widely used current collector material in electrochemical systems due to its three-dimensional porous structure, high surface area, light weight, mechanical strength, chemical stability, and excellent electrical and thermal conductivity. It features an interconnected pillar network that forms small (~50 μm) and large (~500 μm) open pores, providing effective pathways for electrolyte transport and ensuring high surface accessibility. The surface of nickel foam naturally contains metallic nickel (Ni^0) and its oxidized forms (Ni^{2+}), such as nickel oxide (NiO) and nickel hydroxide ($Ni(OH)_2$) [24]. Period functionalization, the native oxide layer on the Ni foam surface is typically removed using concentrated acid to expose a clean metallic surface. Common methods reported for Ni foam include various methods, each offering distinct advantages in terms of morphology control, composition tunability, and scalability.

3.1.3 Copper foam

Copper foam exhibits rich redox chemistry and excellent electrical conductivity. Over the past decade, researchers have increasingly leveraged the highly porous surface of copper foam for the fabrication of various nanomaterials. Beyond serving as a structural support, copper foam can also act as a precursor for other copper-based derivatives, making it a highly versatile material platform for advanced functional materials. Among the available fabrication techniques, electrodeposition remains the most widely adopted method due to its cost-effectiveness, ease of electrolyte preparation, and precise control over film thickness. Copper foams typically possess a uniformly distributed three-dimensional (3D) porous network with ultrahigh porosity (>95%) and significant specific tensile strength. These structures are generally fabricated through a combination of pretreatment steps, chemical or electrochemical deposition, calcination, and thermal reduction. Recent research efforts have focused on enhancing the surface characteristics of copper foam using advanced modification techniques to increase porosity and improve electrochemical performance [25].

3.2 Carbon foam

Carbon foam with microporous structures is a material composed of various carbon forms, such as fibers, tubes, spheres, flakes, or granules, each possessing unique porosity characteristics. Typically, the macropores in carbon foams are interconnected, forming an open-cell structure that results in a high geometric surface area. This distinctive structure is directly linked to the fundamental properties of carbon materials, including low density, high thermal stability, hydrophobic surfaces, and excellent thermal and electrical conductivity. Reticulated vitreous carbon (RVC) foams were first utilized as electrodes, benefiting from their honeycomb-like structure that supported electrochemical applications [26]. Since then, carbon foams have continued to attract interest and have been researched and developed for a wide range of applications. These include heat energy storage both as containers and thermal conductivity enhancers, electrodes in electrochemical systems, gas adsorption and separation using molecular sieve structures, oil spill absorption and recovery, and electromagnetic wave reflection. These diverse applications highlight the significant engineering potential and value of carbon foams [27].

Carbon foams can be produced through a variety of techniques. However, in recent years, the synthesis of carbon foams from polymers and biomass has attracted significant attention due to their recyclability and environmentally friendly characteristics. One widely used synthesis method involves thermosetting materials combined with a sacrificial template, such as nickel foam. This process typically includes the deposition of carbon *via* chemical vapor deposition (CVD) onto the foam structural template, followed by acid washing to remove the residue. Nevertheless, controlling the pore structure at the nanometer scale remains a significant challenge for this method. Alternatively, melamine foams or biomass materials with high nitrogen content have emerged as promising options. These materials can serve as both porous templates and carbon sources in the direct carbonization process, leading to carbon foams with enhanced structural integrity and electrical conductivity. Nitrogen doping not only improves the electrical conductivity of the carbon framework but also introduces defect sites at the molecular level, which play a crucial role in the electrocatalytic performance of the material, particularly when used as an electrode in electrochemical reaction systems [28].

3.3 Synthesis of an electrocatalyst on foam substrate materials

3.3.1 Hydrothermal

This method is the most widely adopted approach for the direct growth of nanostructured catalysts. This technique enables facile, low-cost, and binder-free fabrication of well-adhered active materials with controllable chemical composition and morphology. Through hydrothermal reactions, various transition metal oxides, sulfides, phosphides, and hydroxides can be synthesized directly onto NF, yielding electrodes with enhanced electrochemical performance due to the strong interfacial bonding and efficient charge transport pathways afforded by the 3D architecture [29]. This technique allows for precise control over the morphology and physical properties of the catalyst at relatively low temperatures compared to conventional high-temperature methods. One limitation of this approach is the use of surfactants, which can be difficult to remove and may negatively affect the catalytic performance [8].

3.3.2 Electrodeposition

Electrodeposition is a method used to synthesize electrocatalysts for various energy conversion applications. This involves the deposition of materials onto a conductive substrate using an electric current. There are several types of electrodeposition techniques, including galvanostatic electrodeposition (constant current), potentiostatic electrodeposition (constant potential), and co-electrodepositiontion. The process typically uses a three-electrode system consisting of a working electrode, counter electrode, and reference electrode. The current density, electrolyte concentration, pH, and temperature can affect the electrodeposition process and the properties of the resulting electrocatalyst. Electrodeposition enables the synthesis of porous nanomaterials with high specific surface areas, which can enhance electrocatalytic activity [30].

Electrodeposition method has been extensively employed to deposit thin catalyst layers such as transition metal hydroxides, layered double hydroxides (LDHs), and noble metals onto NF substrates. This method allows precise control over catalyst loading and thickness under ambient conditions, making it suitable for scalable electrode fabrication [31–33]. In the electrochemical method, metal ions from a precursor solution are reduced and deposited onto the copper foam surface under an applied potential, resulting in the formation of thin metal films or composite layers [34]. Electrodeposition is favored due to its simplicity, fast processing time, and control of the thickness and structure *via* current or time [35].

3.3.3 Chemical vapor deposition (CVD)

Chemical vapor deposition (CVD) is a widely used technique for growing 2D materials, including transition metal dichalcogenides (TMDs). Some key points about the CVD method for growing 2D TMDs are as follows: it allows for the controllable synthesis of large-scale, high-quality 2D TMD materials. Several key factors affect the CVD growth process, including the temperature, carrier gas, growth time, and precursor ratio. The CVD method has been used to successfully grow various 2D TMD materials, such as MoS_2, WS_2, $MoSe_2$, and WSe_2 nanosheets [36]. Proper design of the growth system and choice of growth substrates are crucial for obtaining wafer-scale monolayer TMD films and millimeter-sized single crystals. The CVD method has also been extended to growing more complex structures, such as TMD alloys, heterostructures, and superlattices. The CVD method can control the growth of thin film layers through several key factors, such as temperature, as higher temperatures generally promote better crystallinity and larger domain sizes. The type and flow rate of the carrier gas influence the transport of the precursors and can affect the shape and quality of the grown crystals. For example, the introduction of H_2 gas can optimize the flake shape. The precursor ratio and feeding rate affect the growth rate, domain size, and composition of the thin films. Moreover, longer growth times can lead to larger crystal sizes by providing a continuous precursor supply [37].

3.3.4 In situ oxidation

This method involves the release of copper ions from the foam into an alkaline solution, where they react with a strong oxidizing agent to form copper hydroxide ($Cu(OH)_2$) under mild conditions. The resulting $Cu(OH)_2$ can be further converted into other functional copper-based compounds, such as copper oxide (CuO), through controlled thermal treatment [38, 39].

Among the four synthesis methods, chemical vapor deposition (CVD) offers the highest level of precision, allowing for atomic-scale control over film thickness, crystallinity, composition, and morphology. Electrodeposition also provides high precision, especially in controlling catalyst thickness, loading, and structure by adjusting deposition parameters such as current, time, and electrolyte composition. Hydrothermal synthesis enables moderate control over morphology and chemical composition but is less precise due to the influence of surfactants and batch variability. *In situ* oxidation provides the least control, as it largely depends on the intrinsic properties of the foam substrate and mild reaction conditions, limiting the tunability of the final catalyst structure.

4. Foam substrates for electrocatalytic water splitting (HER and OER)

Foam substrates such as nickel, copper, and carbon foams provide a unique three-dimensional (3D) conductive scaffold with large surface area and mechanical stability, making them highly suitable for hosting bifunctional electrocatalysts capable of catalyzing both hydrogen evolution (HER) and oxygen evolution (OER) reactions. The use of these porous frameworks promotes gas diffusion, enhances electron transport, and allows effective anchoring of nanostructured catalysts, thus improving overall water-splitting efficiency.

4.1 Nickel foam

Nickel foams have been advantageously utilized as electrode substrates because they present a high active surface area and high conductivity [40]. Both HER and OER are widely used because of their excellent conductivity and mechanical strength. Relative to other transition metals such as Fe and Co, Ni alloys and their composites exhibit excellent resistance to corrosion in alkaline media, unlike Ni metal alone, which leads to relatively low catalytic activity [41]. Transition metal phosphides such as CoP and Ni_2P, directly grown on nickel foam, have shown outstanding performance. For example, CoP/NF demonstrated an overpotential of just 45 mV, outperforming commercial Pt/C [42]. For OER, Layered double hydroxides (LDHs) are a class of two-dimensional materials composed of divalent and trivalent metal cations (M^{2+} and M^{3+}, respectively) coordinated with hydroxide anions and charge-compensating interlayer anions. However, despite their promising catalytic behavior, the practical application of LDHs is limited by their inherently low electrical conductivity and relatively modest specific surface area (SAs). To overcome these limitations, recent efforts have focused on the direct growth of LDH architectures on conductive, high-surface-area substrates. Among these, the integration of LDHs with NF has been demonstrated to significantly enhance the OER performance. NiFe-layered double hydroxides (NiFe-LDH) integrated on NF exhibit low onset overpotentials (~230 mV) and long-term stability [43]. Further improvements have been achieved by incorporating elements like Co or Mn to create NiCoFe or NiMnFe LDH, which can deliver even higher current densities [44]. Additionally, the catalysts, such as $NiCo_2S_4$/NF and NiFe-PBA/NF, demonstrate high activity in both HER and OER, enabling efficient overall water splitting. In one study, a NiFe-LDH/Ni/$NiCo_2S_4$/NF catalyst delivered an overall water-splitting voltage of 1.53 V [45].

4.2 Copper foam

Copper foam, while less commonly used than nickel foam, has emerged as a promising substrate due to its high porosity and ease of functionalization. Nanostructured MoS_2 based alloys on Cu foam exhibit overpotentials 218 mV for HER, benefiting from good electron conductivity and catalyst-support interaction [46]. Cu foam has been modified with protective layers such as NiCo-$NiCoO_2$ nanoheterostructures or conductive polymers. These coatings serve as physical barriers and chemically stable interfaces that inhibit oxidation, dissolution of the underlying copper, and strong mechanical adhesion/interaction between conductive substrate, thereby improving long-term durability and maintaining electrocatalytic performance during HER [47]. Although Cu foam is unstable under the high anodic potentials required for OER, especially in alkaline media, where copper species (Cu, Cu_2O, and CuO) can corrode, dissolve, or oxidize into soluble forms such as $Cu(OH)_4^{2-}$.Less stable under oxidative conditions. However, some studies have reported Cu foam with high performance for OER and HER as a bifunctional catalyst. By engineering hybrid architectures, such as NiMnOx nanosheets grown onto CuOx nanowires, Cu foam can support high-performance bifunctional electrocatalysts with enhanced durability and reduced overpotentials. Owning to the formation of intimate interfacial contact and valence interactions between the Cu and Ni/Mn species can lead to improved charge transport and partial stabilization of the underlying CuOx during OER [48].

4.3 Carbon foam

Carbon foams, including graphite and biomass-derived carbon, are attractive due to their light weight, chemical stability, and tunable porosity. Compared to metal foams like nickel or copper, carbon foams generally exhibit lower electrical conductivity but offer significant advantages in terms of cost-effectiveness and corrosion resistance, especially in acidic or neutral pH environments, due to their light weight, chemical stability, and tunable porosity. Metal nanoparticles like Ni or Co embedded on carbon foam show low overpotential. By forming an interconnected conductive network with graphene, carbon foam improves electron transfer kinetics, reducing the chances of corrosion caused by inefficient charge flow. Its hierarchical porous structure enhances mass transfer, minimizing localized conditions that could lead to corrosion. Together, these features create a durable environment that protects the active sites and improves the long-term stability of the electrocatalyst [49]. For OER, carbon foam modified with metal oxides or LDHs can provide competitive performance. The synergy between conductive carbon and active metal centers promotes oxygen evolution, especially when dopants like nitrogen (N) or phosphite (P) are introduced. Carbon foam's open framework facilitates electrolyte access and bubble release. However, its hydrophobicity must often be tuned to maintain optimal wettability and mass transport [50].

As mentioned above, by comparison to the typical foam substrate, nickel foam remains the most widely used substrate due to its superior stability and compatibility with alkaline electrolytes. However, copper and carbon foams present unique opportunities for customized architectures and cost reduction. The future of foam-based electrocatalysis lies in the rational design that combines multiple materials, fine-tunes surface chemistry, and leverages advanced fabrication methods such as electrodeposition or 3D printing.

5. Foam substrates for electrochemical CO_2 reduction reaction

Copper foam plays a crucial role in the electrochemical reduction of CO_2 reduction reaction (CO_2RR) because of its electronic and structural properties. Although pure bulk copper has limited selectivity at lower overpotentials, nanostructured Cu, especially in the form of electrodeposited foam, has demonstrated notable activity in converting CO_2 to products such as CO, formate, and hydrocarbons. This is primarily attributed to their unique surface morphology, which includes dendritic structures with high roughness and large electrochemically active surface areas. Such structures expose a greater number of low-coordinated atoms (e.g., edges, steps, and kinks), which are known to stabilize reaction intermediates such as *COOH and facilitate CO_2 adsorption and activation. Copper foam served as the base material for synthesizing bimetallic AgCu foams through hydrogen-bubble-templated electrodeposition. Even before incorporating silver, the porous Cu foam exhibited measurable CO_2RR activity, particularly at lower overpotentials. CO and formate were produced, demonstrating that the inherent catalytic properties of copper remain effective when nanostructured [51].

Beyond the surface area, the intrinsic electrical conductivity of the Cu foam was selected to ensure efficient electron transport during electrolysis. The Cu foam was first oxidized to grow Cu nanowires, creating a high-surface-area scaffold that facilitated the uniform electrodeposition of Zn nanosheets and a thin indium layer on the scaffold (as shown in **Figure 2A**). These modifications led to the formation of a nanostructured InZnCu catalyst, which exhibited excellent CO_2-to-CO conversion with a Faradaic efficiency over 90%. This is crucial for reducing the overall energy input of the CO_2RR process. Furthermore, its mechanical stability supports long-term operation, which is a requirement for any system aimed at practical or industrial-scale CO_2 conversion. For the OER, Cu foam was used again, which was modified with copper hydroxide nanowires and coated with NiCo layered double hydroxide (LDH). This combination creates a robust and conductive anode material that facilitates efficient water oxidation, operates at low overpotentials, and exhibits high durability. In essence, Cu foam acts as a versatile platform that can be engineered to support high-performance electrocatalysts on both the cathode and anode sides of the CO_2RR-OER system, as shown in **Figure 2B**, which shows the low energy requirement to convert CO_2 to CO per ton [52].

Cu foam substrate electrodes typically suffer from moderate selectivity, surface reconstruction, and stability issues during prolonged operations. In contrast, recent research has focused on silver (Ag)-based foam electrodes. The nano-Ag foam in this study is not a conventional foam but a hierarchical nanosheet structure grown *in situ* on a porous Ag skeleton. This design substantially increases the electrochemically active surface area (ECSA). In the nano-Ag foam system, the catalyst and support were unified, leading to superior mechanical stability and electronic conductivity throughout the electrode structure [53].

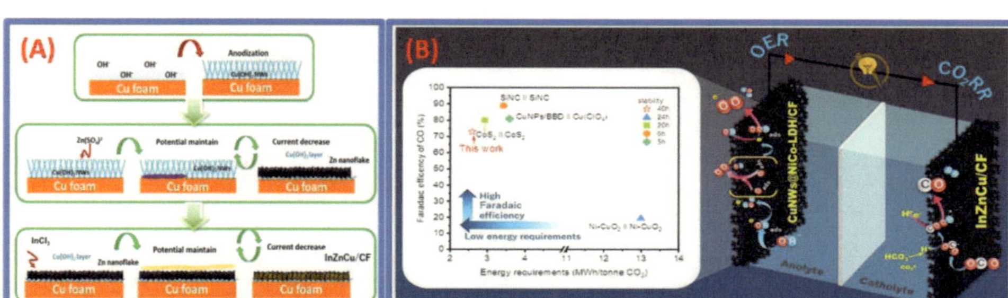

Figure 2.
(A) Schematic of the synthesis of the InZnCu base copper foam electrode. (B) Overall CO_2RR-OER performance of an assembled InZnCu/CF (cathode)||Cu(OH)2 NWs@NiCo-LDH (anode). Reproduced with permission from Ref. [52].

Method	Key features	Advantages	Disadvatages	Materials
Hydrothermal	Growth *via* reactions in a sealed vessel at moderate temperatures.	Low cost, surfactant-free (optional), binder-free, good adhesion, and strong interfacial bonding.	Surfactants may be hard to remove and can affect catalytic performance.	Transition metal oxides, sulfides, phosphides, hydroxides, and carbon foam
Electrodeposition	Deposition using electrical current (e.g., galvanostatic, potentiostatic, and o-electrodeposition).	Fast, scalable, ambient conditions, precise control over loading and thickness, and simple setup.	Sensitive to parameters like pH, electrolyte, current density, and temperature.	Metal hydroxides, LDHs, noble metals, composite layers
Chemical vapor deposition (CVD)	Gas-phase growth of materials on a substrate, often at high temperatures.	High quality, uniform, wafer-scale coatings, control over structure/morphology.	Requires complex setup, high temps, time-consuming, and not ideal for all substrates.	2D TMDs (MoS_2, WS_2, $MoSe_2$, and WSe_2), alloys, heterostructures, carbon foam
In situ oxidation	Oxidation of metal foam in alkaline solution to form hydroxides/oxides.	Simple, mild conditions, self-sourced material from foam, easy conversion to other copper-based materials.	Limited to metal foam substrates (like Cu), may need post-treatment (e.g., thermal conversion).	$Cu(OH)_2$, CuO, other copper-based compounds

Table 1.
Comparison the method of synthesis electrocatalyst on foam substrate materials.

DOI: http://dx.doi.org/10.5772/intechopen.1012509

Therefore, foam-based electrodes, particularly those using nickel, copper, and carbon foam substrates, demonstrate promising performance for water splitting and CO_2 reduction reactions. As **Table 1** shows, an overview of foam-based materials presents various catalysts that show bifunctional activity for both hydrogen evolution reaction (HER) and oxygen evolution reaction (OER), enabling efficient overall water splitting. Common synthesis methods include hydrothermal growth, electrodeposition, and phosphorization to create nanostructured active materials on the foam substrates. Transition metal-based materials like NiFe-layered double hydroxides (LDHs), cobalt phosphides, and nickel selenides are frequently used as active catalysts. Low overpotentials (often <200 mV) are achieved for HER and OER, indicating high catalytic activity. For CO_2 reduction, copper-based foams modified with other metals like Ag, In, Zn show selectivity for different products like CO, formate, and hydrocarbons. Hierarchical nanostructures combining different materials (e.g., $NiSe_2$-$NiMoO_4$) often show enhanced performance compared to single-component catalysts. The 3D porous structure of foams provides high surface area and facilitates mass transport, contributing to the high activity. Further optimization of composition, morphology and synergistic effects between components could lead to even better performing catalysts for these important electrochemical reactions (**Table 2**).

Electrode	Synthesis method	Application	Performance	References
$NiFe/NiCo_2O_4$/Ni Foam	Hydrothermal, Electrodeposition	HER, OER	HER 105 mV at 10 mA/cm^2 OER 340 mV at 1200 mA/cm^2	[54]
NiCoP nanosheet arrays/Ni Foam	Hydrothermal, phosphorization	HER, OER	HER 133 mV at 50 mA/cm^2 OER 308 mV at 50 mA/cm^2	[55]
MoP/Ni_2P/NF	Hydrothermal, phosphorization	Overall water splitting (two-electrode system)	1.55 V at 10 mA/cm^2	[56]
$NiCo_2S_4$@NiFe LDH/NF	Hydrothermal	Overall water splitting (two-electrode system)	1.6 V	[57]
$Ni_{0.75}Fe_{0.125}V_{0.125}$-LDHs/ NF	Hydrothermal	Overall water splitting (two-electrode system)	1.591 V	[58]
$NiSe_2$-$NiMoO_4$/NF	Hydrothermal and selenization	Overall water splitting (two-electrode system)	1.37 V at 10 mA/cm^2	[59]
CoFe@NiFe/NF	Hydrothermal and electrodeposition	Overall water splitting (two-electrode system)	1.59 V at 10 mA/cm^2	[60]
NiFe/Cu nanowires on Cu foam	Chemical oxidation, Thermal treatment and electrodeposition	Overall water plitting (two-electrode system)	1.54 V at 100 mA/cm^2	[61]
Cu/Cu_2O foam	Electrodeposition, Electroless plating	HER	HER 290 mV	[62]

Electrode	Synthesis method	Application	Performance	References
Cu foam@BiNW	*In situ* electrochemical transformation, Thermal treatment	CO_2RR	$FE_{formate}$ 95%	[63]
PEI-Sn/Cu foam	Electrodeposition and impregnation	CO_2RR	FE_{CO} 92.3%	[64]
InZnCu/CF	Electrodeposition	CO_2RR	FE_{HCOOH} 93.7%	[52]
$CoSe_2$/ carbon foam	Inkjet printing, pyrolysis, and selenization	HER, OER	HER 122.6 mV at 10 mA/cm^2 OER 400 mV at 10 mA/cm^2	[65]

Table 2.
Overview of various foam-based electrode materials, their synthesis methods, and applications.

6. Key design strategies in foam electrodes for application

6.1 Structure-property relationships

Recent studies have significantly advanced the quantitative understanding of the relationship between the pore structures of foam electrodes and their electrochemical performance. Liu et al. developed nickel foam electrodes with hierarchical pore distributions, comprising macropores (>500 μm) and micropores (<50 μm), and quantitatively demonstrated, *via* electrochemical impedance spectroscopy (EIS), that macropores enhance gas bubble release, while micropores promote electrolyte infiltration. This hierarchical structuring effectively reduced mass transport resistance by approximately 30% at high current densities (≥200 mA cm^{-2}), increasing the HER current density by about 25% compared to conventional foams Weber et al. utilized selective laser melting (SLM) additive manufacturing to precisely tailor pore structures with controlled overall porosity (50–70%) and pore size distribution (10–100 μm). Their study revealed that foam electrodes with macroscopic porosity above 65% and a specific surface area of approximately 12 m^2 g^{-1} exhibited a notable enhancement (20%) in CO Faradaic efficiency at −1.0 V vs. RHE for CO_2RR, while effectively suppressing competing HER side reactions [66].

In another innovative approach, Hoffmann et al. applied dynamic hydrogen-bubble templating (DHBT) to fabricate hierarchical silver foams containing cellular macropores (300–600 μm) interconnected with dense nested micropores (1–10 μm). Quantitative X-ray computed tomography (XCT) and scanning electron microscopy (SEM) analyses confirmed that macropores facilitated rapid gas release, while the microporous network significantly enhanced the adsorption of reaction intermediates ($^*CO_2$, *COOH), leading to a CO Faradaic efficiency as high as 98% and improved current density (~50 mA cm^{-2}) at −0.8 V vs. RHE [67]. Shi et al. further demonstrated the significance of multi-scale structuring by integrating Bi_2O_3 nanosheets onto copper foam surfaces with additional silane-based hydrophobic modification, creating hierarchical macro-micro-nano interfaces. Their electrode achieved over 90% selectivity for formate ($HCOO^-$) at −0.9 V_RHE, maintaining exceptional stability with less than 5% Faradaic efficiency loss after 50 hours [68].

Addressing the trade-off between high porosity and mechanical stability, Hu et al. introduced an elastic PDMS phase into the carbon foam skeleton, creating secondary pore gradients. This approach maintained high porosity (~85%) while restoring compressive modulus to about 92% of the original foam. As a result, the modified foam electrode displayed excellent mechanical robustness, with only a 3% degradation in performance after 1000 hours of oxygen evolution reaction (OER) testing at a current density of 100 mA cm^{-2} [69]. Similarly, Qiu et al. explored hierarchical porous transport layers (PTLs) for proton exchange membrane water electrolyzers (PEMWE), applying a fine titanium-powder layer (20 μm) onto a coarse macroporous substrate (200 μm). This engineered porosity optimized two-phase transport, significantly reducing combined ohmic and concentration polarization by 12% at 2 A cm^{-2}, and improving energy efficiency by 8% [70].

Lastly, single-atom catalyst (SAC) integration into porous electrodes has emerged as a promising strategy. Zhu et al. synthesized indium single-atom catalysts anchored on carbon foams *via* a NaCl-templating method, achieving a controlled pore size distribution (50–200 μm). Through X-ray photoelectron spectroscopy (XPS) and density functional theory (DFT) studies, they demonstrated that the synergistic interaction between In-N_4 single-atom sites and the pore structures greatly enhanced CO_2 adsorption and stabilized reaction intermediates, achieving an exceptional formate Faradaic efficiency of 82% and a stable current density of 67 mA cm^{-2} at −1.4 V vs. RHE over 21 hours [71].

In summary, recent advances in hierarchical pore structuring, surface functionalization, and mechanical reinforcement have enabled precise balancing among pore structure, interfacial chemistry, and mechanical robustness. These developments provide comprehensive guidelines for optimizing foam electrodes to achieve enhanced efficiency, product selectivity, and durability in HER, OER, and CO_2RR applications.

6.2 Gas-bubble effect

During the hydrogen evolution (HER) and oxygen evolution reactions (OER), gas bubbles frequently adhere to the electrode surface, forming a "bubble shield" that obstructs active sites and hinders ion transport. Consequently, the performance of foam electrodes in water electrolysis and CO_2 reduction reactions (CO_2RR) is highly contingent upon effective gas-bubble management. In conventional foam electrodes, gas bubbles accumulate within the porous structure, obstructing catalytic sites and disrupting ion transport at the solid-liquid interface. This results in increased overpotentials and diminished reaction efficiency. Notably, the hierarchical micro/nanostructure significantly enhances the electrochemical surface area (ECSA), thereby exposing more active sites for catalysis. These surfaces exhibit underwater superaerophobicity, contributing to a more discontinuous three-phase contact line. This phenomenon suppresses bubble aggregation, promotes the formation of smaller bubbles, and enhances water-splitting performance. The low-adhesive nanosheet arrays on the surface play a crucial role in reducing bubble size and increasing the frequency of bubble release. This feature, in conjunction with the underlying microscale porous structure, augments mass transfer at the electrode/electrolyte interface. The high bubble contact angle of the superaerophobic surface further facilitates bubble detachment [72]. Additionally, the combination of microscale pores and groove patterns induces an asymmetric Laplace pressure, which drives the lateral spreading, transport, and timely detachment of bubbles from the surface. The 3D nanosheet arrays (≈30 nm thick and ≈2 μm long) further increase the surface area, contributing to higher reaction rates. Collectively, these micro/nanostructures help establish a balanced solid-liquid-gas microenvironment, reduce gas-bubble adhesion, and disrupt the three-phase contact line, collectively improving gas release dynamics and catalytic efficiency.

7. Recent advances and future outlook

In recent years, significant breakthroughs have been achieved in the research on foam electrodes, particularly concerning multifunctional design and system integration. Notably, additive manufacturing-based 3D printing techniques facilitate the fabrication of microarchitected foam electrodes from metallic or carbon-based materials, allowing customizable pore size distributions, gradient porosity, and precision-engineered conductive networks. These advancements not only greatly enhance the specific surface area (up to 20–30 $m^2\ g^{-1}$) and electrical conductivity (>10^4 S cm^{-1}) but also maintain the exceptional mechanical robustness of the material. The resulting electrodes outperform conventional foam counterparts in both HER and OER performance under high-current-density operation, while also establishing a foundational platform for integrating multiphysics-coupled functionalities such as sensing and energy storage [73]. Second, the engineering of interfaces by introducing oxygen functional groups or applying metal oxide coatings onto the surfaces of graphene foam significantly enhances the wettability of the electrolyte and optimizes the energetics of intermediate adsorption. This synergy effectively reduces the overpotentials to less than 40 mV for the HER and less than 200 mV for the OER, while also extending operational lifetimes beyond 500 hours, thereby demonstrating a robust potential for operation in extreme environments [74]. In CO_2RR applications, electrodeposited In-Zn alloy catalytic layers on Cu foam substrates utilize nanostructure modulation and control of the local alkaline microenvironment to achieve an FE of CO exceeding 85% with over 100 hours of current retention. This configuration enhances productivity by nearly 50% compared to planar Cu electrodes [52]. Recent advances in room-temperature flow CO_2 electrolyzers have elevated their energy efficiency to 55%, concurrently validating the feasibility of large-area modular deployment [75]. However, the scalable production of foam electrodes continues to face challenges in terms of process reproducibility, consistency of microstructural characteristics between batches, and effective cost management. Future efforts should prioritize the development of low-cost, self-supporting composite foam architectures and ensure their deep integration with ion-exchange membranes and electrochemical stacks. By employing modular and flexible designs alongside real-time monitoring, multifunctional foam electrodes can be synergistically utilized in industrial-scale applications such as water splitting, CO_2 valorization, and renewable energy storage.

8. Conclusions

Foam-based substrate materials are increasingly utilized in electrochemical applications within the realm of renewable energy, particularly in water electrolysis and CO_2 reduction. These technologies are pivotal in addressing climate change and achieving carbon neutrality goals. Foam electrodes present substantial advantages over planar electrodes due to their three-dimensional porous architecture, which offers increased surface area, enhanced mass transport, and efficient gas bubble release, thereby improving reaction rates and catalytic efficiency. Metal foams, such as those composed of nickel and copper, exhibit high conductivity and can function as both current carriers and catalytic centers. Nickel foam is particularly prevalent due to its superior conductivity and mechanical strength. Conversely, copper foam is advantageous in terms of cost-effectiveness and corrosion resistance, especially in acidic or neutral environments, and can be modified with various catalysts, such as

DOI: http://dx.doi.org/10.5772/intechopen.1012509

NiFe-layered double hydroxides (LDHs), to enhance performance. The reduction of CO_2 is particularly noteworthy due to the unique electronic and structural properties of nanostructured copper foam, which has demonstrated significant activity in converting CO_2 into CO, formate, and hydrocarbons. Carbon foams, which can support metal nanocatalysts, offer high conductivity and corrosion resistance at a reduced cost, making them advantageous in terms of cost-effectiveness and corrosion resistance, particularly in acidic or neutral environments. This chapter underscores two principal design strategies: (i) structure-property relationships, where factors such as pore size, surface area, conductivity, and mechanical stability critically influence electrochemical performance and (ii) gas-bubble management, where effective gas bubble release is essential for maintaining high catalytic activity. Hierarchical micro/nanostructures can facilitate bubble detachment and enhance overall efficiency. Nonetheless, challenges persist in scalable production, including process reproducibility and cost management. Future research should prioritize the development of low cost, self-supporting composite foam architectures and their integration with ion-exchange membranes and electrochemical stacks.

Author details

Suchada Sirisomboonchai[1], Rongchi Dai[1]* and Koyo Norinaga[2]

1 Department of Chemical Systems Engineering, Graduate School of Engineering, Nagoya University, Aichi, Japan

2 Institutes of Innovation for Future Society, Nagoya University, Nagoya, Aichi, Japan

*Address all correspondence to: dairongchi1994@gmail.com

References

[1] Nitopi S, Bertheussen E, Scott SB, Liu X, Engstfeld AK, Horch S, et al. Progress and perspectives of electrochemical CO(2) reduction on copper in aqueous electrolyte. Chemical Reviews. 2019;**119**:7610-7672. DOI: 10.1021/acs.chemrev.8b00705

[2] Khalil M, Kadja GTM, Nugroho FAA, Sutanto LG, Jiwanti PK, Abdi FF, et al. Suppressing the competing hydrogen evolution reaction in CO_2 electroreduction: A review. Renewable and Sustainable Energy Reviews. 2024;**206**:114869. DOI: 10.1016/j.rser.2024.114869

[3] He H, Zhang Q, Wang Z, Pan S, Zhao Y, Zhang X. Advances and practical prospects for bias-free photovoltaic-driven electrochemical water splitting systems. Advanced Energy Materials. 2024;**14**:2303713. DOI: 10.1002/aenm.202303713

[4] Pan F, Yang X, O'Carroll T, Li H, Chen KJ, Wu G. Carbon catalysts for electrochemical CO_2 reduction toward multicarbon products. Advanced Energy Materials. 2022;**12**:2200586. DOI: 10.1002/aenm.202200586

[5] Kim C, Talapaneni SN, Dai L. Porous carbon materials for CO_2 capture, storage and electrochemical conversion. Materials Reports: Energy. 2023;**3**:100199. DOI: 10.1016/j.matre.2023.100199

[6] Meng W, Pang R, Li M, Han L, Kong X, Zhang D, et al. Integrated catalyst-substrate electrodes for electrochemical water splitting: A review on dimensional engineering strategy. Small. 2024;**21**:e2310469. DOI: 10.1002/smll.202310469

[7] Zheng W, Liu M, Lee LYS. Best practices in using foam-type electrodes for electrocatalytic performance benchmark. ACS Energy Letters. 2020;**5**:3260-3264. DOI: 10.1021/acsenergylett.0c01958

[8] Sun H, Kim H, Song S, Jung W. Copper foam-derived electrodes as efficient electrocatalysts for conventional and hybrid water electrolysis. Materials Reports: Energy. 2022;**2**:100092. DOI: 10.1016/j.matre.2022.100092

[9] Girichandran N, Saedy S, Kortlever R. Electrochemical CO_2 reduction on a copper foam electrode at elevated pressures. Chemical Engineering Journal. 2024;**487**:150478. DOI: 10.1016/j.cej.2024.150478

[10] Liu F, Wang F, Hao X, Fan Z, Tan J. Effects of foam cathode electrode structure on alkaline water electrolysis for hydrogen production. Chemical Engineering Science. 2024;**298**:120307. DOI: 10.1016/j.ces.2024.120307

[11] Feidenhans'l AA, Regmi YN, Wei C, Xia D, Kibsgaard J, King LA. Precious metal free hydrogen evolution catalyst design and application. Chemical Reviews. 2024;**124**:5617-5667. DOI: 10.1021/acs.chemrev.3c00712

[12] Su G, Yang J, Yin J. Advancements in electrocatalysts for oxygen evolution reaction: A review of catalysts in acidic media. ChemElectroChem. 2025;**12**:2400559. DOI: 10.1002/celc.202400559

[13] Vadakkayil A, Dunlap-Shohl WA, Joy M, Bloom BP, Waldeck DH. Improved catalyst performance for the oxygen evolution reaction under a chiral bias.

ACS Catalysis. 2024;**14**:17303-17309. DOI: 10.1021/acscatal.4c04477

[14] Wiranarongkorn K, Eamsiri K, Chen Y-S, Arpornwichanop A. A comprehensive review of electrochemical reduction of CO_2 to methanol: Technical and design aspects. Journal of CO2 Utilization. 2023;**71**:102477. DOI: 10.1016/j.jcou.2023.102477

[15] Wang F, Wang B, Zhang T, Liu J. Electrochemical CO_2 reduction to C3+ products: A review. Green Carbon. 2025;2;9. DOI: 10.1016/j.greenca.2025.02.009

[16] Zhan C, Dattila F, Rettenmaier C, Herzog A, Herran M, Wagner T, et al. Key intermediates and Cu active sites for CO(2) electroreduction to ethylene and ethanol. Nature Energy. 2024;**9**:1485-1496. DOI: 10.1038/s41560-024-01633-4

[17] Bhattacharya A, Calmidi VV, Mahajan RL. Thermophysical properties of high porosity metal foams. International Journal of Heat and Mass Transfer. 2002;**45**:1017-1031. DOI: 10.1016/S0017-9310(01)00220-4

[18] Shen T, Wang P, Shi F, Xu P, Zhang G. Metal foam-based functional materials application in advanced oxidation and reduction processes for water remediation: Design. Mechanisms, and Prospects, Chemical Engineering Journal. 2024;**500**:156825. DOI: 10.1016/j.cej.2024.156825

[19] Lefebvre L-P, Banhart J, Dunand DC. Porous metals and metallic foams: Current status and recent developments. Advanced Engineering Materials. 2008;**10**:775-787. DOI: 10.1002/adem.200800241

[20] Fu W, Li Y. Fabrication, processing, properties, and applications of closed-cell aluminum foams: A review. Materials. 2024;**17**:560

[21] Tan WC, Saw LH, Thiam HS, Xuan J, Cai Z, Yew MC. Overview of porous media/metal foam application in fuel cells and solar power systems. Renewable and Sustainable Energy Reviews. 2018;**96**:181-197. DOI: 10.1016/j.rser.2018.07.032

[22] Du R, Jin X, Hübner R, Fan X, Hu Y, Eychmüller A. Engineering self-supported noble metal foams toward electrocatalysis and beyond. Advanced Energy Materials. 2020;**10**:1901945. DOI: 10.1002/aenm.201901945

[23] Su Z, Chen T. Porous noble metal electrocatalysts: Synthesis, performance, and development. Small. 2021;**17**:2005354. DOI: 10.1002/smll.202005354

[24] Grdeń M, Alsabet M, Jerkiewicz G. Surface science and electrochemical analysis of nickel foams. ACS Applied Materials and Interfaces. 2012;**4**:3012-3021. DOI: 10.1021/am300380m

[25] Das M, Biswas A, Purkait T, Boruah T, Bhardwaj S, Das SK, et al. The versatility of the dynamic hydrogen bubble template derived copper foam on the emerging energy applications: Progress and future prospects. Journal of Materials Chemistry A. 2022;**10**:13589-13624. DOI: 10.1039/D2TA01355J

[26] Edwards NS, Conley JC, Reichenberger MA, Nelson KA, Tiner CN, Hinson NJ, et al. Characterization of reticulated vitreous carbon foam using a frisch-grid parallel-plate ionization chamber. Nuclear Instruments and Methods in Physics Research Section A: Accelerators, Spectrometers, Detectors and Associated Equipment. 2018;**892**:93-97. DOI: 10.1016/j.nima.2018.03.005

[27] Inagaki M, Qiu J, Guo Q. Carbon foam: Preparation and application. Carbon. 2015;**87**:128-152. DOI: 10.1016/j.carbon.2015.02.021

[28] Wang R, Yu J, Islam F, Tahmasebi A, Lee S, Chen Y. State-of-the-art research and applications of carbon foam composite materials as electrodes for high-capacity lithium batteries. Energy and Fuels. 2020;**34**:7935-7954. DOI: 10.1021/acs.energyfuels.0c01802

[29] Xiao L, Wang X, Fu L, Deng W, Liang B, Li R, et al. Zn-Co needle-grass-shaped nanostructure in situ-grown on porous nickle foam as an electrocatalyst for efficient HER. Journal of Electronic Materials. 2025;**54**:5501-5510. DOI: 10.1007/s11664-025-11914-7

[30] Kale MB, Borse RA, Mohamed AGA, Wang Y. Electrocatalysts by electrodeposition: Recent advances, synthesis methods, and applications in energy conversion. Advanced Functional Materials. 2021;**31**:2101313. DOI: 10.1002/adfm.202101313

[31] Poimenidis IA, Papakosta N, Klini A, Farsari M, Konsolakis M, Loukakos PA, et al. Electrodeposited Ni foam electrodes for increased hydrogen production in alkaline electrolysis. Fuel. 2023;**342**:127798. DOI: 10.1016/j.fuel.2023.127798

[32] Yang J, Lian L, Ruan H, Xie F, Wei M. Nanostructured porous MnO_2 on Ni foam substrate with a high mass loading via a CV electrodeposition route for supercapacitor application. Electrochimica Acta. 2014;**136**:189-194. DOI: 10.1016/j.electacta.2014.05.074

[33] Cai Z-X, Song X-H, Wang Y-R, Chen X. Electrodeposition-assisted synthesis of Ni2P Nanosheets on 3D graphene/Ni foam electrode and its performance for electrocatalytic hydrogen production. ChemElectroChem. 2015;**2**:1665-1671. DOI: 10.1002/celc.201500239

[34] Eugénio S, Silva TM, Carmezim MJ, Duarte RG, Montemor MF. Electrodeposition and characterization of nickel–copper metallic foams for application as electrodes for supercapacitors. Journal of Applied Electrochemistry. 2014;**44**:455-465. DOI: 10.1007/s10800-013-0646-y

[35] Klingan K, Kottakkat T, Jovanov ZP, Jiang S, Pasquini C, Scholten F, et al. Reactivity determinants in electrodeposited Cu foams for electrochemical CO_2 reduction. ChemSusChem. 2018;**11**:3449-3459. DOI: 10.1002/cssc.201801582

[36] Fruehwald HM, Liebsch Y, Javed U, Lebius H, Grygiel C, Rahali R, et al. Basal-plane pores activate monolayer MoS2 for the hydrogen evolution reaction. ACS Catalysis. 2025;**15**:3768-3776. DOI: 10.1021/acscatal.4c07970

[37] Zhang Y, Yao Y, Sendeku MG, Yin L, Zhan X, Wang F, et al. Recent Progress in CVD growth of 2D transition metal dichalcogenides and related heterostructures. Advanced Materials. 2019;**31**:e1901694. DOI: 10.1002/adma.201901694

[38] Liu Y, Cao X, Jiang D, Jia D, Liu J. Hierarchical CuO nanorod arrays in situ generated on three-dimensional copper foam via cyclic voltammetry oxidation for high-performance supercapacitors. Journal of Materials Chemistry A. 2018;**6**:10474-10483. DOI: 10.1039/C8TA00945G

[39] Li H, Liu Y, He F, Yang H, Li Z, Zhou Q, et al. In situ grown Cu-based metal-organic framework on copper foam as high-performance electrocatalysts for oxygen evolution

reaction. International Journal of Hydrogen Energy. 2020;**45**:21540-21546. DOI: 10.1016/j.ijhydene.2020.05.234

[40] Chaudhari NK, Jin H, Kim B, Lee K. Nanostructured materials on 3D nickel foam as electrocatalysts for water splitting. Nanoscale. 2017;**9**:12231-12247. DOI: 10.1039/C7NR04187J

[41] Cechanaviciute IA, Antony RP, Krysiak OA, Quast T, Dieckhofer S, Saddeler S, et al. Scalable synthesis of multi-metal electrocatalyst powders and electrodes and their application for oxygen evolution and water splitting. Angewandte Chemie (International Ed. in English). 2023;**62**:e202218493. DOI: 10.1002/anie.202218493

[42] Cao S, You N, Wei L, Huang C, Fan X, Shi K, et al. CoP microscale prism-like superstructure arrays on Ni foam as an efficient bifunctional electrocatalyst for overall water splitting. Inorganic Chemistry. 2020;**59**:8522-8531. DOI: 10.1021/acs.inorgchem.0c00959

[43] Lu Z, Xu W, Zhu W, Yang Q, Lei X, Liu J, et al. Three-dimensional NiFe layered double hydroxide film for high-efficiency oxygen evolution reaction. Chemical Communications (Cambridge). 2014;**50**:6479-6482. DOI: 10.1039/c4cc01625d

[44] Yang Q, Li T, Lu Z, Sun X, Liu J. Hierarchical construction of an ultrathin layered double hydroxide nanoarray for highly-efficient oxygen evolution reaction. Nanoscale. 2014;**6**:11789-11794. DOI: 10.1039/c4nr03371j

[45] Gong XS, Liu X, Zhou J. Unique hierarchical NiFe-LDH/Ni/NiCo(2) S(4) heterostructure arrays on nickel foam for the improvement of overall water splitting activity. Nanoscale. 2025;**17**:5301-5315. DOI: 10.1039/d4nr05248j

[46] Tian Y, Qi C, Zhou R, Li D, Han T. Photoelectrocatalytic hydrogen evolution reaction stimulated by the surface plasmon resonance effect of copper and silver surrounded with MoS(2). RSC Advances. 2023;**13**:25246-25252. DOI: 10.1039/d3ra04357f

[47] Qazi UY, Javaid R, Zahid M, Tahir N, Afzal A, Lin X-M. Bimetallic NiCo–NiCoO2 nano-heterostructures embedded on copper foam as a self-supported bifunctional electrode for water oxidation and hydrogen production in alkaline media. International Journal of Hydrogen Energy. 2021;**46**:18936-18948. DOI: 10.1016/j.ijhydene.2021.03.046

[48] Sirisomboonchai S, Li X, Kitiphatpiboon N, Channoo R, Li S, Ma Y, et al. Fabrication of CuOx nanowires@NiMnOx nanosheets core@shell-type electrocatalysts: Crucial roles of defect modification and valence states for overall water electrolysis. Journal of Materials Chemistry A. 2020;**8**:16463-16476. DOI: 10.1039/d0ta04172f

[49] Wei D, Chen L, Tian L, Ramakrishna S, Ji D. Melamine-derived carbon foam-supported graphene–CoNi nanocomposites as high-performance OER/HER bifunctional electrocatalysts. Journal of Chemical Technology and Biotechnology. 2022;**98**:910-918. DOI: 10.1002/jctb.7294

[50] Zhang R, He S, Lu Y, Chen W. Fe, Co, N-functionalized carbon nanotubes in situ grown on 3D porous N-doped carbon foams as a noble metal-free catalyst for oxygen reduction. Journal of Materials Chemistry A. 2015;**3**:3559-3567. DOI: 10.1039/c4ta05735j

[51] Kottakkat T, Klingan K, Jiang S, Jovanov ZP, Davies VH, El-Nagar GAM, et al. Electrodeposited AgCu foam catalysts for enhanced reduction of

CO(2) to CO. ACS Applied Materials and Interfaces. 2019;**11**:14734-14744. DOI: 10.1021/acsami.8b22071

[52] Sirisomboonchai S, Machida H, Bao Tran KV, Kawasumi M, Norinaga K. Efficient CO_2 electrochemical reduction by a robust electrocatalyst fabricated by electrodeposition of indium and zinc over copper foam. ACS Applied Energy Materials. 2022;**5**:9846-9857. DOI: 10.1021/acsaem.2c01564

[53] Zhang Y, Wang Y, Li J, Zhang L, Zhu X, Fu Q, et al. Highly selective CO(2) electroreduction in an exsolution-induced flow cell using a hierarchical monolithic nano-Ag foam electrode. Nanoscale. 2025;**17**:11605-11614. DOI: 10.1039/d5nr00181a

[54] Xiao C, Li Y, Lu X, Zhao C. Bifunctional porous NiFe/$NiCo_2O_4$/Ni foam electrodes with triple hierarchy and double synergies for efficient whole cell water splitting. Advanced Functional Materials. 2016;**26**:3515-3523. DOI: 10.1002/adfm.201505302

[55] Li Y, Zhang H, Jiang M, Kuang Y, Sun X, Duan X. Ternary NiCoP nanosheet arrays: An excellent bifunctional catalyst for alkaline overall water splitting. Nano Research. 2016;**9**:2251-2259. DOI: 10.1007/s12274-016-1112-z

[56] Du C, Shang M, Mao J, Song W. Hierarchical MoP/Ni2P heterostructures on nickel foam for efficient water splitting. Journal of Materials Chemistry A. 2017;**5**:15940-15949. DOI: 10.1039/C7TA03669H

[57] Liu J, Wang J, Zhang B, Ruan Y, Lv L, Ji X, et al. Hierarchical $NiCo_2S_4$@NiFe LDH heterostructures supported on nickel foam for enhanced overall-water-splitting activity. ACS Applied Materials and Interfaces. 2017;**9**:15364-15372. DOI: 10.1021/acsami.7b00019

[58] Dinh KN, Zheng P, Dai Z, Zhang Y, Dangol R, Zheng Y, et al. Ultrathin porous NiFeV ternary layer hydroxide nanosheets as a highly efficient bifunctional electrocatalyst for overall water splitting. Small. 2018;**14**:1703257. DOI: 10.1002/smll.201703257

[59] Yu T, Xu Q, Chen J, Qian G, Zhuo X, Yang H, et al. Boosting urea-assisted water splitting by constructing sphere-flower-like NiSe2-NiMoO4 heterostructure. Chemical Engineering Journal. 2022;**449**:137791. DOI: 10.1016/j.cej.2022.137791

[60] Yang R, Zhou Y, Xing Y, Li D, Jiang D, Chen M, et al. Synergistic coupling of CoFe-LDH arrays with NiFe-LDH nanosheet for highly efficient overall water splitting in alkaline media. Applied Catalysis B: Environmental. 2019;**253**:131-139. DOI: 10.1016/j.apcatb.2019.04.054

[61] Yu L, Zhou H, Sun J, Qin F, Yu F, Bao J, et al. Cu nanowires shelled with NiFe layered double hydroxide nanosheets as bifunctional electrocatalysts for overall water splitting. Energy and Environmental Science. 2017;**10**:1820-1827. DOI: 10.1039/C7EE01571B

[62] Farahbakhsh N, Sanjabi S. Activated Cu/Cu2O foam with Ni nanoparticles for electrocatalytic activity enhancement of hydrogen evolution reaction (HER) in acidic media. Journal of Industrial and Engineering Chemistry. 2019;**70**:211-225. DOI: 10.1016/j.jiec.2018.10.018

[63] Zhang X, Sun X, Guo S-X, Bond AM, Zhang J. Formation of lattice-dislocated bismuth nanowires on copper foam for enhanced electrocatalytic CO_2 reduction at low overpotential. Energy and Environmental Science. 2019;**12**:1334-1340. DOI: 10.1039/C9EE00018F

[64] Yan Y, Peng Y, Song Y, Wang R, Wang H, Bian Z. Polyethyleneimine-reinforced Sn/Cu foam dendritic self-supporting catalytic cathode for CO_2 reduction to HCOOH. Chemosphere. 2022;**301**:134704. DOI: 10.1016/j.chemosphere.2022.134704

[65] Wang Y, Yu J, Liu Q, Liu J, Chen R, Zhu J, et al. Porous carbon foam loaded CoSe2 nanoparticles based on inkjet-printing technology as self-supporting electrodes for efficient water splitting. Electrochimica Acta. 2023;**438**:141594. DOI: 10.1016/j.electacta.2022.141594

[66] Weber N, Linkhorst J, Keller R, Wessling M. Tailoring pore networks – Gas diffusion electrodes via additive manufacturing. Advanced Materials Technologies. 2023;**8**:2300720. DOI: 10.1002/admt.202300720

[67] Hoffmann H, Paulisch-Rinke MC, Gernhard M, Jannsch Y, Timm J, Brandmeir C, et al. Multi-scale morphology characterization of hierarchically porous silver foam electrodes for electrochemical CO(2) reduction. Communications Chemistry. 2023;**6**:50. DOI: 10.1038/s42004-023-00847-z

[68] Shi T, Liu D, Liu N, Zhang Y, Feng H, Li Q. Triple-phase Interface engineered hierarchical porous electrode for CO(2) Electroreduction to Formate. Advanced Science (Weinh). 2022;**9**:e2204472. DOI: 10.1002/advs.202204472

[69] Hu H, Wang J, Li C, Li L. Reinforcement strategies to improve the mechanical properties of nanofibrous aerogels: A review. ACS Applied Nano Materials. 2024;**7**:26546-26578. DOI: 10.1021/acsanm.4c03465

[70] Qiu C, Xu Z, Chen F-Y, Wang H. Anode engineering for proton exchange membrane water electrolyzers. ACS Catalysis. 2024;**14**:921-954. DOI: 10.1021/acscatal.3c05162

[71] Zhu X, Xu Y, Ran L, Chen S, Qiu X. Three-dimensional porous indium single-atom catalysts with improved accessibility for CO(2) reduction to formate. Inorganic Chemistry. 2024;**63**:3893-3900. DOI: 10.1021/acs.inorgchem.3c04273

[72] Li Y, Li K, Li L, Gao J, Wang Z, Zou W, et al. Bubble-guidance breaking gas shield for highly efficient overall water splitting. Advanced Materials. 2024;**36**:e2405493. DOI: 10.1002/adma.202405493

[73] Scheideler WJ, Im J. Recent advances in 3D printed electrodes - bridging the Nano to mesoscale. Advanced Science (Weinh). 2025;**12**:e2411951. DOI: 10.1002/advs.202411951

[74] Zhao M, Jin H, Chen Z, Fang R, Feng H, He D. Interface engineering of scalable graphene foam induces hydrophilic/aerophobic trimetallic (Fe, Co, Ni) nanosheets for efficient oxygen evolution. Surfaces and Interfaces. 2025;**62**:106082. DOI: 10.1016/j.surfin.2025.106082

[75] Sajna MS, Zavahir S, Popelka A, Kasak P, Al-Sharshani A, Onwusogh U, et al. Electrochemical system design for CO_2 conversion: A comprehensive review. Journal of Environmental Chemical Engineering. 2023;**11**:110467. DOI: 10.1016/j.jece.2023.110467

Chapter 5

Numerical Study of a Direct-Irradiated Solar-Driven Pyrolysis Reactor

Yuhan Jin, Yibo Wu, Xinyi Chen, Youwei Yang, Yaning Zhang and Ruming Pan

Abstract

Solar energy, as a renewable energy source currently attracting much attention in the energy field, is of great significance for its effective utilization and conversion for environmental protection and energy industry development. This study proposes a solar-driven photothermal reactor and establishes a mathematical model coupled with multi-physical fields. Specifically, the radiative transfer of the numerical model takes into account different wavelengths to improve the simulation accuracy. Further, the effects of operating condition parameters such as xenon lamp power, inlet flow rate, and foam's porosity and pore diameter on the photothermal field performance in the reactor were comparatively analyzed. The results showed that lower xenon lamp power, higher inlet flow rate, and smaller foam's porosity and pore diameter could improve the photothermal (fluid) energy conversion rate of the solar reactor. The findings contribute to diversifying solar energy utilization strategies and support future efforts to optimize reactor energy conversion performance.

Keywords: solar-driven reactor, radiative heat transfer, foams, multi-physical field coupling, numerical simulation

1. Introduction

Energy has always been a critical material foundation for the development, progress, and survival of human society. As societal development advances, energy demand continues to rise. Fossil fuels remain the dominant source of primary energy consumption, primarily owing to their high energy density and convenient transportability [1, 2]. However, these carbon-based fossil fuels generate substantial amounts of air pollutants and greenhouse gases, negatively impacting ecosystems and the climate. This results in challenges including global warming, ecological degradation, and energy scarcity, which hinder sustainable development. To address the increasingly severe environmental and energy challenges, vigorously developing renewable green energy has become an inevitable trend in the energy sector. Renewable resources such as solar, wind, and biomass present considerable potential for further advancement. Notably, solar energy is widely regarded as a leading candidate among future energy sources owing to its abundance, environmental friendliness, renewability, and sustainability [3]. Solar energy, as both the direct and indirect source of all terrestrial energy, is extensively researched and harnessed [4]. Among all renewable resources, solar energy is the most abundant, with yearly availability far outweighing humanity's overall energy requirements [5]. How to effectively absorb, store, and convert it is a critical issue in solar energy utilization.

In solar thermochemistry, the solar reactor is a key part that can realize the purpose of converting solar energy to heat and then to chemical energy [6–8]. Since the last century, numerous research teams have been designing and developing solar reactors for hydrogen from methane reforming. However, experimental results indicate that these reactors suffer from several issues, including localized high temperatures, poor thermal stability, and significant heat loss. Consequently, many researchers have focused on optimizing solar reactors through improvements in optical design, geometric configuration, heat transfer models, and solution methods.

Recent studies have concentrated on enhancing photothermal performance and conversion efficiency by modifying reactor structures. For instance, Bai et al. [9] applied a spectrally selective coating ($TiON_x$) to the outer layer of the reactor and added vacuum tubes to increase solar radiation absorption while reducing energy loss due to convection and conduction. This structural modification significantly improved light absorption and thermal insulation, resulting in a 32.9% increase in solar-fuel conversion efficiency. Jiang et al. [10] investigated the transfer characteristics of radiation intensity using different Fresnel lens shapes, designing three optimized Fresnel lens structures via the Monte Carlo method. The results revealed that the use of Fresnel lenses and parabolic mirrors not only regulated the photothermal performance of the reactor but also optimized the solar collector-reactor coupled concentrator system, leading to cost reductions and reasonable thermal storage efficiency. Additionally, Bala Chandran et al. [11, 12] designed a novel tubular flow channel that integrates radiative absorption with foam catalyst filling, achieving a 12% increase in fuel productivity and over 90% heat recovery efficiency. This design allowed for continuous fuel production under sunlight through alternating redox steps across six separate reactors within a half-cycle. The tube filled with alumina mesh porous ceramics outside the reactor cavity served as an efficient heat recovery device, effectively increasing reaction temperature and overall thermal efficiency. From these studies, it is evident that structural modifications to solar reactors can significantly enhance their performance.

Moreover, foam-filled solar reactor is the research hotspot. The complex pore structure favors the transmission of radiation in its interior, and the high specific surface area enhances convective heat transfer between the fluid and solid phases, making it an efficient medium for thermal exchange. Guene et al. [13] investigated the radiative transfer and heat transfer processes in a solar receiver with a quartz glass cover and in foam filled with porous materials. The results indicated that the temperature and heat flux of the diffuse radiating surface were strongly affected by the heat transfer coefficient, which led to considerable heat loss near the solar flux inlet of the receiver. Du et al. [14] developed an optimization approach integrating genetic algorithms with thermal analysis for solar collectors containing porous structures. Both single-objective and multi-objective optimizations were performed to enhance receiver performance. The results of single-objective optimization revealed that greater porosity and elevated inlet velocity contribute to improved thermal performance of solar receivers. The multi-objective optimization approach seeks to achieve a balance between minimizing flow resistance and maximizing heat transfer efficiency in the receiver. In conclusion, optimizing heat transfer and flow within the solar reactor is crucial for improving energy conversion efficiency.

To gain deeper insights into the reactor's heat and mass transfer processes and to explore how various factors influence the photothermal performance, numerical simulation methods combined with experimental validation offer an efficient and cost-effective approach. Zhang et al. [15] numerically investigated the thermal and flow behaviors in a high-temperature solar thermochemical reactor incorporating foams, considering various thermophysical property models. The study highlighted the necessity of using both LTNE and radiative transfer models to effectively assess thermal characteristics in high-temperature thermochemical processes. Tong et al. [16] employed direct pore-scale numerical simulations to explore the coupled heat transfer behavior in a porous volumetric solar receiver, based on a model developed at the pore scale. The results showed that a higher inlet velocity enhanced convective heat transfer between the liquid and solid phases, significantly reducing overall temperature, and that greater porosity can minimize reflection loss. Thus, optimizing the inlet velocity and foam scale parameters helps reduce heat loss and improve thermal efficiency.

In summary, to address challenges such as local overheating, large heat losses, and poor thermal stability in solar reactors, efficiency can be improved by designing the optical structure and modifying the reactor's design. Exploring heat and radiation transfer in the reactor helps improve efficiency by lowering associated losses. Additionally, numerical simulations can be used to optimize influencing factors. This study proposes a novel solar-driven photothermal reactor. A numerical model with multi-physical field coupling, including the radiation transfer process at different wavelengths, is established. The influencing factors of temperature distribution, radiation distribution, and energy utilization inside the reactor are analyzed to advance the reactor's capability in converting solar input into thermal energy.

2. Methodology

2.1 Physical model

Figure 1 shows the simplified physical models of the solar simulator and solar reactor. The solar simulator is composed of a xenon lamp source and a concentrating mirror. The solar reactor is composed of an alumina insulation cavity, a quartz glass

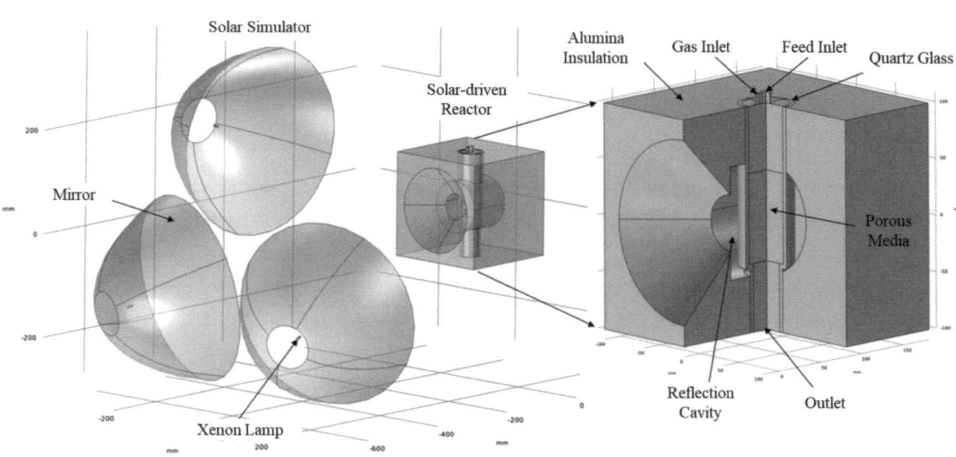

Figure 1.
The physical model in numerical simulation.

tube, and foams [17]. The xenon lamp source is defined as a composite of spherical and cylindrical surfaces. The condenser mirror is modeled as an ellipsoid. The xenon lamp of the solar simulator generates light, which is guided into the reactor via reflective mirrors. The foam absorbs part of the incoming light directly, and the rest is delivered through the reflection chamber. The foam is made of alumina ceramic material, with a diameter of 42 mm and a height of 80 mm. Inert gas flows into the reactor through the inlet, ensuring a chemically stable environment. Entering through the feed inlet, the reactants are thermally decomposed into volatile compounds by the foam and subsequently discharged from the outlet. The model assumes that the xenon lamp source behaves as a Lambertian source on a set surface. The concentrating mirror exhibits specular reflection properties, while the various surfaces inside the reactor exhibit diffuse reflection and emission properties for radiation in both short and long wavelength ranges. The glass's selective transmittance and absorption processes are approximated by assuming partial absorption and transmittance for short-wave radiation, based on the absorption factor. For long-wave radiation, the glass is considered opaque with a defined diffuse emissivity. It is concluded that the radiation transmission process in the foam can be treated as a linear superposition. The short-wave and long-wave radiation transmission processes are considered separately as gray medium processes. Regarding fluid dynamics and thermal transport within the reactor, the fluid flow is approximated as compressible ideal-fluid laminar flow [18]. The thermal transport in both fluid and solid domains is modeled based on Fourier's law, combined with the governing equations of continuity, energy, and momentum conservation [19].

2.2 Governing equations

This study analyzed radiative heat transfer, involving short-wavelength solar radiation (0–2 μm) and long-wavelength infrared emission (2–20 μm) [17]. The short-wavelength radiative transfer using the xenon lamp as the radiation source is expressed as follows:

$$I_{\lambda_1,i}\left(\vec{\Omega_j}\right) = \sum_k BRDF_{\lambda_1,i}\left(\vec{\Omega_k},\vec{\Omega_j}\right)\cos\left\langle \vec{n}_i,\vec{\Omega_k}\right\rangle I_{\lambda_1,l}\left(\vec{\Omega_k}\right)\Delta\Omega_k\Big|_{\overrightarrow{A_lA_i}=\vec{\Omega_k}} = I_{E,\lambda_1,i}\left(T_i,\vec{\Omega_j}\right) \quad (1)$$

$$Q_{\mathrm{rad},i} = \Delta A_i \sum\nolimits_{j,k}\left[\alpha_{\lambda_1,i}\left(\vec{\Omega_j}\right)G_{\lambda_1,i}\left(\vec{\Omega_j}\right) - E_{\lambda_1,i}\left(T_i,\vec{\Omega_j}\right)\right]\Delta\Omega_j\Delta\lambda_k \quad (2)$$

where λ_1 denotes the short wavelength, $\vec{\Omega}$ specifies the direction of the solid angle, $\Delta\Omega$ is the size of differential elements, $I_{\lambda_1,i}$ is the element i's effective radiation intensity in the short wavelength λ_1, $BRDF_{\lambda_1,i}$ is the bidirectional reflectance distribution function, $\vec{n}_i$ denotes the surface normal vector, $I_{E,\lambda_1,i}$ represents the element i's emissive intensity in the short wavelength λ_1, T_i is the uniform temperature, $Q_{\mathrm{rad},i}$ corresponds to the radiative heat source of element i, ΔA_i indicates the surface area element., $\alpha_{\lambda_1,i}$ represents the element i's absorptance in the short wavelength λ_1, $G_{\lambda_1,i}$ is the xenon lamp irradiation power, and $E_{\lambda_1,i}$ is the emission radiation intensity in short wavelength [17, 20].

The P1 approximation model is adopted to evaluate radiative heat transfer within the foam.

$$\nabla \cdot (D_{\lambda_1,\text{Pl}} \nabla G_{\lambda_1}) - \kappa_{\lambda_1,\text{a}} G_{\lambda_1} = 0 \tag{3}$$

$$\nabla \cdot (D_{\lambda_2,\text{Pl}} \nabla G_{\lambda_2}) - \kappa_{\lambda_2,\text{a}} (G_{\lambda_2} - 4\pi I_\text{b}) = 0 \tag{4}$$

where D_{Pl} denotes the P1 diffusion coefficient, G represents the incident radiation [17], κ_a denotes the absorption coefficient, and I_b represents the blackbody radiation.

The expression for the P1 diffusion coefficient D_{Pl} can be expressed as follows:

$$D_{\lambda_k,\text{Pl}} = \frac{1}{3\left(\kappa_{\lambda_k,a} + \kappa_{\lambda_k,s}\right)} \tag{5}$$

where λ_1 refers to the short wavelength, λ_2 refers to the long wavelength, and κ_s is the scattering coefficient [17]. The absorption coefficient κ_a, scattering coefficient κ_s, and extinction coefficient κ_e are calculated by [17]

$$\kappa_{\lambda_k,\text{a}} = \frac{3\varepsilon_{\lambda_k}(1-\varphi)}{2d_\text{p}} \tag{6}$$

$$\kappa_{\lambda_k,s} = \frac{3\left(2-\varepsilon_{\lambda_k}\right)(1-\varphi)}{2d_\text{p}} \tag{7}$$

$$\kappa_{\lambda_k,\text{e}} = \kappa_{\lambda_k,\text{a}} + \kappa_{\lambda_k,\text{s}} = \frac{3(1-\varphi)}{d_\text{p}} \tag{8}$$

where ε_{λ_k} denotes the emissivity, φ represents the porosity, and d_p corresponds to the pore diameter [17], which is calculated by $d_\text{p} = 0.028PPI^{-0.721}$ (*PPI* indicates pore per inch).

Accordingly, the foam's radiative heat sources are expressed separately for short wavelength $Q_{\lambda_1,r}$ and long wavelength $Q_{\lambda_2,r}$:

$$Q_{\lambda_1,r} = \kappa_{\lambda_1,\text{a}} G_{\lambda_1} \tag{9}$$

$$Q_{\lambda_2,r} = \kappa_{\lambda_2,\text{a}} (G_{\lambda_2} - 4\pi I_\text{b}) \tag{10}$$

Note that the continuity, momentum, and energy equations in the reactor are not provided due to page limitation [21].

3. Results

3.1 Simulation results

Figure 2 demonstrates the temperature distribution under the conditions of a 2.0 kW xenon source, a feed rate of 200 cm^3/min, a porosity of 0.8, and a foam with 10 PPI. The solid temperature in the foams along the radial direction gradually decreases, while the fluid temperature first rises sharply to the peak and then gradually decreases. The formation of the temperature peak in the foam boundary area is due to the local thermal non-equilibrium effect. The reduced fluid flow rate in the boundary area leads to weakened convective heat transfer, and the fluid cannot fully absorb the heat conducted by the solid. Since the solid has a higher thermal conductivity and the fluid phase has a lower thermal conductivity, heat diffuses quickly in the solid and transfers slowly in the fluid, resulting in a significant temperature

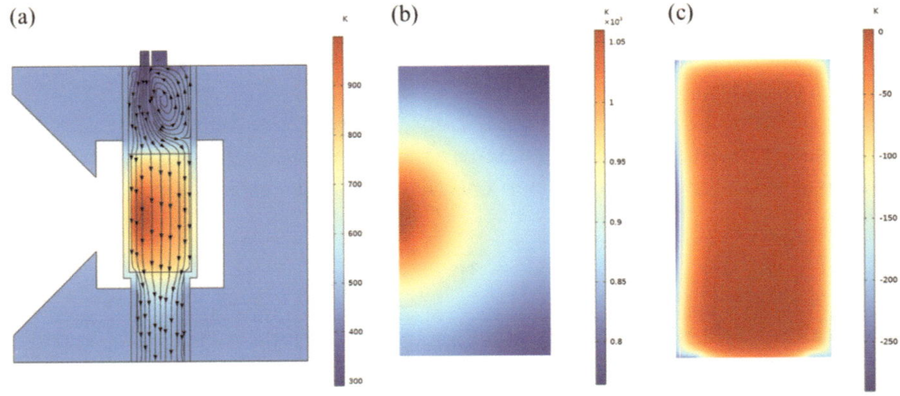

Figure 2.
(a) Temperature distributions Tg at the cross-section of the reactor; (b) solid temperature Ts in foams; (c) solid-fluid temperature difference Ts-Tg in foams.

difference in the boundary area. As the fluid continues to absorb heat, its temperature gradually increases radially, reaching thermal equilibrium with the solid at the peak position, and then the temperature change trends of the two phases tend to be consistent. At the same time, the intensity of short-wave incident radiation decreases significantly, while the intensity of long-wave incident radiation has a smaller trend, and the intensity of long-wave radiation is greater than that of short-wave radiation in the back half of the foam region (x > 80 mm). The solid and fluid phases are considered to be in thermal equilibrium throughout the foam, except near the boundaries of the medium.

3.2 Effect of xenon lamp power

Numerical simulations were conducted to evaluate the influence of xenon lamp power variation (2.0–8.0 kW in 2.0 kW increments) on solar reactor performance characteristics under a constant inlet flow rate of 200 cm^3/min. In this range, the average temperature of the solid skeleton in the foam region increased from 869.28 to 1351.10 K, and the average temperature of the fluid rose from 823.92 to 1296.70 K. The thermal profile exhibits a pronounced dependence on the xenon lamp power intensity, as evidenced in **Figure 3(a)**. The temperature increase is due to the higher intensity of incident radiation when the xenon lamp power increases. This leads to more energy being absorbed by the foam and fluid. **Figure 3(b)** shows that the energy absorbed by the foam ranges from 131.91 W at 2.0 kW to 577.35 W at 8.0 kW. The thermal energy uptake by the fluid increases from 31.73 W at 2.0 kW to 52.96 W at 8.0 kW. Additionally, the energy absorbed by the insulation chamber and quartz glass tube increases by 416.60 W and 268.19 W, respectively. However, it is evident that the energy conversion efficiency decreases with increasing xenon lamp power, from 4.07% to 1.55%. Therefore, a lower xenon lamp power is more favorable for improving the energy conversion rate.

3.3 Effect of inlet velocity

Figure 4(a) reveals that the fluid recirculation zone exhibits progressive expansion with increasing inlet flow velocity. **Figure 4(b)** reveals that higher inlet fluid

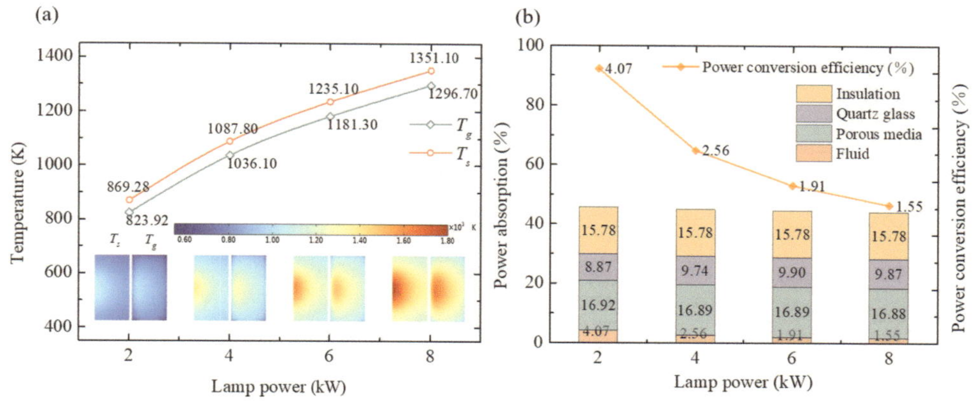

Figure 3.
The average temperature distributions of solid Ts and fluid Tg phases in foams (a) and reactor energy analysis (b) at varying lamp powers (2–8 kW).

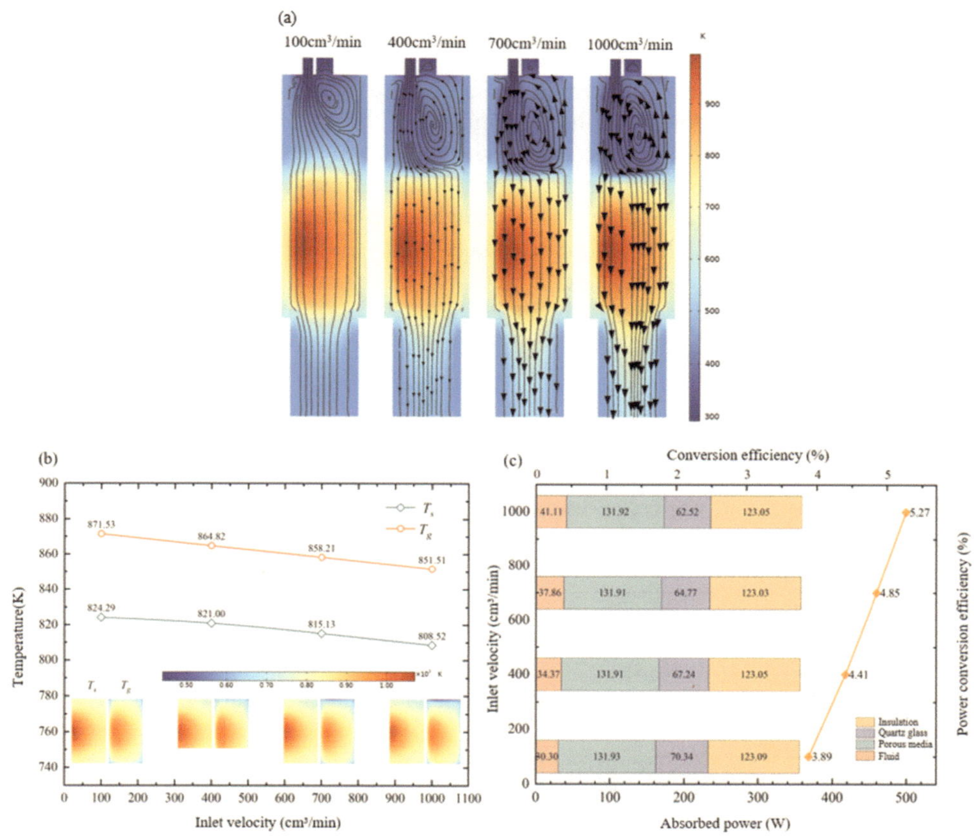

Figure 4.
At different gas inlet velocities: (a) the distribution of gas temperature (Tg) at z = 0; (b) the average solid temperature (Ts) and gas temperature (Tg) within the foams; and (c) the reactor energy analysis.

velocities result in a smaller decrease in the mean solid temperature (Ts) and the mean fluid temperature (Tg). The mean solid temperature decreased from 871.53 to 851.51 K, and the mean fluid temperature decreased from 824.29 to 808.52 K. Due to

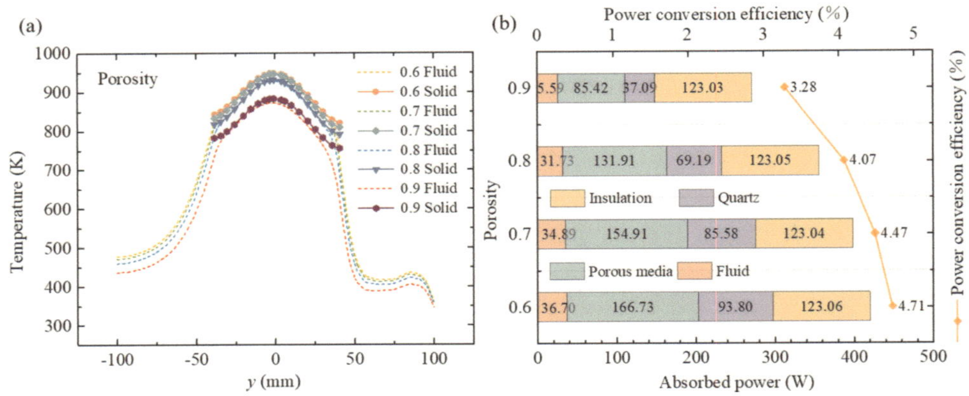

Figure 5.
At different foam porosities: (a) Ts and Tg profiles at the centerline of the reactor and (b) the reactor energy analysis.

the negligible contribution of convective heat transfer relative to conduction and radiation, the average temperature remains largely unchanged. The energy absorbed by each part of the reactor is illustrated in **Figure 4(c)**. As the inlet flow rate rises from 100 to 1000 cm^3/min, the enhanced interfacial heat transfer coefficient facilitates greater energy exchange from the solid structure to the flowing medium inside the foam. The observed increase in fluid temperature beneath the foam toward the reactor outlet is due to this mechanism. Meanwhile, the energy absorbed by the quartz glass tube decreases by 7.82 W, while the energy absorbed by the foam and the holding chamber remains nearly unchanged. The energy conversion efficiency exhibits an upward trend, rising from 3.89% to 5.27%, as the inlet fluid flow rate increases. Therefore, higher inlet fluid velocities are more favorable for improving the energy conversion rate.

3.4 Effect of porosity

Figure 5(a) exhibits the centerline temperature profiles of both fluid and solid phases in the reactor under different porosity conditions. As the porosity increases, the fluid and solid temperatures at the centerline decrease. Specifically, the fluid temperature peak drops from 948.48 to 871.90 K, and the solid temperature peak decreases from 948.79 to 883.07 K. This is attributed to the diminished heat absorption by the porous structure and fluid. **Figure 5(b)** presents the energy distribution inside the reactor. As the porosity increases, the energy absorbed by the fluid, foam, and quartz glass decreases. Notably, the energy absorbed by the foam drops from 166.73 to 85.42 W. This decline arises from the decreased solid fraction at elevated porosities, reducing the reabsorbed radiative energy within the porous structure. As a result, the energy taken up by fluid declines from 36.70 to 25.59 W. Additionally, the energy conversion efficiency drops from 4.71% to 3.28% as porosity increases from 0.6 to 0.9. Therefore, smaller porosity is more favorable for energy conversion.

3.5 Effects of pore size

Figure 6(a) presents the temperature profiles of fluid and solid phases along the reactor centerline for different pore sizes. As pore size decreases, the fluid

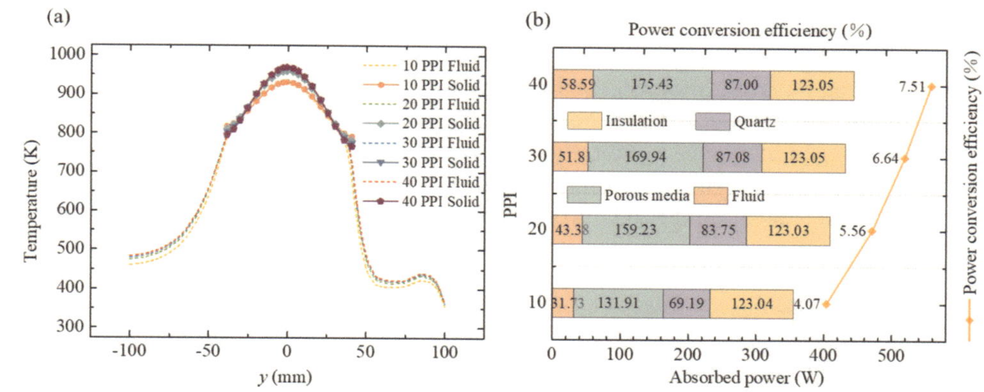

Figure 6.
At different foam pore sizes: (a) Ts and Tg profiles at the centerline of the reactor and (b) the reactor energy analysis.

temperature near the inlet and outlet regions rises. However, the fluid temperature at the inlet remains similar across different pore sizes. In the central region of the foams, both the fluid and solid temperatures are significantly influenced by pore size. As pore size increases, both temperatures decrease, and the temperature distribution becomes more uneven. A reduced pore size elevates the extinction coefficient, leading to a larger radiative heat flow gradient and greater temperature inhomogeneity, which can cause localized thermal stresses and deformation in the foam. To reduce the risk of local thermal stress and deformation, the following measures can be taken: optimize pore size and structure; select moderate foam pore size; design gradient pore structure to evenly distribute heat; improve thermal conductivity; use high thermal conductivity materials or thermal conductive coatings to reduce temperature gradients; provide reasonable layout and buffer design; introduce flexible connections or thermal expansion buffers to relieve mechanical stress; predict temperature fields and stress concentration areas through multi-physics field simulation; and optimize structure and material selection in advance.

Figure 6(b) shows the energy distribution in the reactor. The results reveal that the energy absorbed by the fluid, foam, and quartz glass increases significantly with the reduction in pore size from 10 to 40 PPI. The energy absorbed by the foam increased by 43.52 W. Additionally, the energy absorbed by the fluid due to convective heat transfer nearly doubled, from 31.73 to 58.59 W, attributable to the improved interfacial heat transfer coefficient. Furthermore, the optical-fluid energy conversion efficiency increased from 4.07% to 7.51% as the pore size decreased. This indicates that smaller pore sizes are more favorable for improving optical-fluid energy conversion.

4. Conclusion

In this work, a new physical model and a multi-physics coupled mathematical model for a solar-driven photothermal reactor are developed. The distribution of various physical fields in the solar reactor under specific operating conditions is obtained through simulations using COMSOL software. The study investigates how different factors, including xenon lamp power, inlet velocity, foam's porosity, and pore size, affect the temperature distribution of the fluid and foams, the short-wave and long-wave radiation distribution, and energy utilization rate in the reactor. The principal findings emerging from this investigation may be summarized as follows.

A novel solar-driven photothermal reactor is proposed, utilizing a multispectral radiation transfer mechanism. This design reduces radiation and heat losses, enhances heat transfer, and improves photothermal conversion performance. The study examined the impacts of varying operating conditions. The results showed that as xenon lamp power increased, both fluid and porous solid temperatures exhibited substantial elevation. Energy absorption by the fluid, foams, quartz glass, and insulation cavity also increased. However, the energy conversion efficiency decreases with increasing xenon lamp power from 4.07% to 1.55%. Higher inflow velocities diminished thermal energy transfer from the porous solid to the fluid. Increasing the velocity improved the optical-fluid energy conversion rate (3.89–5.27%). Thus, reducing the xenon lamp power and increasing the inlet velocity can improve the photothermal energy conversion rate. A comparative analysis of the impact of different foam structural parameters was conducted. The simulation results revealed that increasing the porosity reduced the fluid and foam solid temperature. The energy absorbed by the fluid, foams, and quartz glass decreased, leading to a reduced optical-fluid energy conversion rate (4.71–3.28%). Smaller pore diameters promoted greater energy absorption by the fluid, foams, and quartz glass, thus improving the light-fluid energy conversion rate (4.07–7.51%). Therefore, the photothermal conversion rate is enhanced by controlling the foam's porosity and pore size.

In this study, a xenon lamp simulator was employed for numerical modeling. Its spectral characteristics in the visible and near-infrared regions closely resemble those of natural sunlight, facilitating systematic analysis under controlled conditions. However, natural solar irradiation exhibits significant fluctuations in both intensity and spectral composition, which may impact the actual performance of solar energy systems. Therefore, the present results primarily reflect the system's response under idealized illumination conditions. To enhance the engineering applicability of the model, future work will involve experimental validation under real solar irradiation scenarios. Subsequent research may focus on the following aspects: (1) incorporating dynamic solar radiation data into numerical simulations to more accurately represent system performance under varying weather conditions; (2) conducting outdoor prototype testing to assess the reactor's adaptability under conditions of intermittent high irradiance and partial shading; and (3) developing system designs with adaptive regulation capabilities to strengthen the stability and overall efficiency of solar energy utilization.

Acknowledgements

This work was supported by the National Key R&D Program of China (Grant No. 2024YFE0116800) and Natural Science Foundation of Heilongjiang Province, China (Grant No. LH2024E040).

Conflict of interest

The authors declare no conflict of interest.

Author details

Yuhan Jin, Yibo Wu, Xinyi Chen, Youwei Yang, Yaning Zhang and Ruming Pan*
School of Energy Science and Engineering, Harbin Institute of Technology, Harbin, China

*Address all correspondence to: ruming.pan@hit.edu.cn

References

[1] Wang Z, Song H, Liu H, et al. Coupling of solar energy and thermal energy for carbon dioxide reduction: Status and prospects. Angewandte Chemie. 2020;**132**(21):8092-8111. DOI: 10.1002/anie.201907443

[2] Gustavsson L, Nguyen T, Sathre R, et al. Climate effects of forestry and substitution of concrete buildings and fossil energy. Renewable and Sustainable Energy Reviews. 2021;**136**:110435. DOI: 10.1016/j.rser.2020.110435

[3] Zhu P, Li H, Zheng P. A hybrid framework for estimating photovoltaic dust content based on UAV hyperspectral images. International Journal of Applied Earth Observation and Geoinformation. 2025;**139**:104500. DOI: 10.1016/j.jag.2025.104500

[4] Spasiano D, Marotta R, Malato S, et al. Solar photocatalysis: Materials, reactors, some commercial, and pre-industrialized applications. A comprehensive approach. Applied Catalysis B: Environmental. 2015; **170-171**:90-123. DOI: 10.1016/j.apcatb.2014.12.050

[5] Anon. Statistical Review of World Energy 2022. London, UK: BP. p. 2022

[6] Hooshdaran B, Haghshenasfard M, Hosseini SH, et al. CFD modeling and experimental validation of biomass fast pyrolysis in a conical spouted bed reactor. Journal of Analytical and Applied Pyrolysis. 2021;**154**:105011. DOI: 10.1016/j.jaap.2020.105011

[7] Diakaridia S, Mario M, Stéphane B, et al. Multi-step kinetic mechanism coupled with CFD modeling of slow pyrolysis of biomass at different heating rates. Chemical Engineering Journal. 2024;**479**:147791. DOI: 10.1016/j.cej.2023.147791

[8] Qi G, Wang Z, Zhang S, et al. Numerical simulation on biomass-pyrolysis and thermal cracking of condensable volatile component. International Journal of Hydrogen Energy. 2020;**45**:12283-12297. DOI: 10.1016/j.ijhydene.2020.02.199

[9] Bai X, Yuan D, Li Y, et al. Ambient sunlight-driven photothermal methanol dehydrogenation for syngas production with 32.9% solar-to-hydrogen conversion efficiency. iScience. 2021; **24**(2):102056. DOI: 10.1016/j.isci.2021.102056

[10] Jiang B, Guene Lougou B, Zhang H, et al. Analysis of high-flux solar irradiation distribution characteristic for solar thermochemical energy storage application. Applied Thermal Engineering. 2020;**181**:115900. DOI: 10.1016/j.applthermaleng.2020.115900

[11] Bala Chandran R, Davidson JH. Model of transport and chemical kinetics in a solar thermochemical reactor to split carbon dioxide. Chemical Engineering Science. 2016;**146**:302-315. DOI: 10.1016/j.ces.2016.03.001

[12] Bala Chandran R, De Smith RM, Davidson JH. Model of an integrated solar thermochemical reactor/reticulated ceramic foam heat exchanger for gas-phase heat recovery. International Journal of Heat and Mass Transfer. 2015; **81**:404-414. DOI: 10.1016/j.ijheatmasstransfer.2014.10.053

[13] Guene Lougou B, Shuai Y, Pan R, et al. Heat transfer and fluid flow analysis of foam solar thermochemical reactor with quartz glass cover. International Journal of Heat and Mass Transfer. 2018;**127**:61-74. DOI: 10.1016/j.ijheatmasstransfer.2018.06.153

[14] Du S, He Y, Yang W, et al. Optimization method for the porous volumetric solar receiver coupling genetic algorithm and heat transfer analysis. International Journal of Heat and Mass Transfer. 2018;**122**:383-390. DOI: 10.1016/j.ijheatmasstransfer.2018.01.120

[15] Zhang H, Guene Lougou B, Pan R, et al. Analysis of thermal transport and fluid flow in high-temperature foams solar thermochemical reactor. Solar Energy. 2018;**173**:814-824. DOI: 10.1016/j.solener.2018.08.015

[16] Du S, Tong Z, Zhang H, et al. Tomography-based determination of Nusselt number correlation for the porous volumetric solar receiver with different geometrical parameters. Renewable Energy. 2019;**135**:711-718. DOI: 10.1016/j.renene.2018.12.001

[17] Pan R, Yang Y, Lougou BG, et al. Thermal performance evaluation of a novel solar-driven pyrolysis reactor. Energy. 2024;**313**:134051. DOI: 10.1016/j.energy.2024.134051

[18] Lougou BG, Wu L, Ma D, et al. Efficient conversion of solar energy through a macroporous ceramic receiver coupling heat transfer and thermochemical reactions. Energy. 2023;**271**:126989. DOI: 10.1016/j.energy.2023.126989

[19] Alobaid F, Almohammed N, Farid MM, et al. Progress in CFD simulations of fluidized beds for chemical and energy process engineering. Progress in Energy and Combustion Science. 2022;**91**:100930. DOI: 10.1016/j.pecs.2021.100930

[20] Wu L, Lougou BG, Jiang B, et al. Experimentally validated numerical model of multi-spectral bands radiative transport in solar receiver/reactor with photo-active porous absorber reacting media. Energy Conversion and Management. 2023;**278**:116740. DOI: 10.1016/j.enconman.2023.116740

[21] Li M, Wang Z, Li J. Numerical study of cold flow characteristics of a new integrated afterburner with strut flame stabilizer. Aerospace Science and Technology. 2024;**153**:109476. DOI: 10.1016/j.ast.2024.109476

Chapter 6

Perspective Chapter: Numerical Investigations on the Regulation of Temperature Distribution in Packed Bed for Exothermic Reactors

Wei Zhang, Yixiong Lin, Koyo Norinaga, Shijie Wu, Yuming Zhang, Zhewen Chen and Jiazhou Li

Abstract

The integration of hydrogen production with CO_2 methanation facilitates the use of existing natural gas infrastructure while lowering CO_2 emissions, thereby removing the necessity for new facilities dedicated to hydrogen transport, storage, and application. However, the CO_2 methanation process is inherently highly exothermic. Elevated reactor temperatures can induce catalyst thermal degradation, sintering, and carbon deposition, which ultimately result in catalyst deactivation and reduced CH_4 production. In this study, numerical simulations were conducted to investigate hot-spot formation and temperature regulation in fixed-bed reactors. The results demonstrated that enhancing catalyst pellet heat transfer significantly suppresses hot spots while maintaining carbon conversion. Furthermore, an optimal packed-bed dilution strategy employing inert particles was identified. This approach can control the hotspot temperature within the operating range of the catalyst, preventing deactivation caused by high temperatures. Additionally, a "CHESS" monolith design was proposed for packed-bed reactors. This design not only sustains high carbon conversion but also enhances heat transfer. The performance of radial flow reactors was explored as well. It was found that the centripetal Z-type radial flow reactor exhibited the best performance. These findings provide valuable insights into temperature management in fixed-bed reactors for CO_2 methanation.

Keywords: packed bed, exothermic reactor, CO_2 methanation, heat management, numerical simulation

1. Introduction

Global warming, long recognized as one of the most critical environmental challenges of this century [1], is substantially attributable to the dramatic rise in atmospheric carbon dioxide concentrations. Atmospheric CO_2 levels, which were around 280 ppm during the pre-industrial period, have surged markedly to 386 ppm by the late twentieth century and have remained elevated to the present day [2]. Against this backdrop, the reduction of atmospheric carbon dioxide concentrations emerges as a matter of paramount urgency. The international community has widely recognized Carbon Capture and Storage (CCS) and Carbon Capture and Utilization (CCU) technologies as indispensable approaches for combating climate change and preserving the planetary environment [3]. Under the current pressing situation, the significance of developing such efficient carbon capture, utilization, and storage technologies in achieving carbon neutrality goals is becoming increasingly pronounced.

CCS technologies have been relatively widely used. Some oil and gas fields utilize CO_2, which is separated from associated gas streams in oil reservoirs, for enhanced oil recovery [4, 5]. For some situations, CO_2 is being reinjected for geological sequestration [6]. However, the implementation of CCS requires extremely high specification equipment, rigorous safety protocols, and significant capital investment [7]. Moreover, geological limitations often restrict its applicability [8]. By contrast, Carbon Capture and Utilization (CCU) technology converts CO_2 into valuable chemicals, demonstrating greater economic viability and long-term potential [9]. Among the emerging CCU processes, CO_2 methanation has gained considerable research attention as a technically attractive alternative [10–14].

Notably, the CO_2 methanation is a highly exothermic process. Elevated reactor temperatures could lead to adverse effects such as catalyst thermal degradation, sintering, and carbon deposition on catalyst particles, thereby reducing catalytic activity and CH_4 yield [15–17]. Thus, controlling the catalyst bed temperature is critical to minimize the risks of carbon deposition and thermal deactivation while maintaining optimal CH_4 production rates [18].

CO methanation can also be utilized for methane synthesis over catalysts, commonly applied in fields such as coal-to-natural gas, coke oven gas, and biomass-derived syngas. Kopyscinski et al. [19] conducted an experimental study employing spatially resolved measurements. They measured the axial gas concentration and temperature profiles within a fluidized bed and analyzed the effects of catalyst loading, gas velocity, and gas composition on the reaction of CO methanation. The results revealed that CO methanation occurs near the inlet region, with a hotspot developing in close proximity where the temperature rise can reach up to 80°C. The CO conversion was found to be virtually 100%. The fluidization quality of the fluidized bed reactor significantly impacts the methanation reaction. Li et al. [20] experimentally demonstrated that the fluidization quality of a Ni/Al_2O_3 catalyst markedly affects the methanation reaction. Because CO methanation is a mole-number-reducing reaction, defluidization readily occurs inside the bed. The addition of alumina effectively improved both the fluidization characteristics and the methanation activity, leading to higher CO conversion and CH_4 yield. Although the CO methanation and CO_2 methanation are both highly exothermic reactions, CO_2 methanation can directly utilize the CO_2 and thus is easier to couple with CO_2 capture technology.

The reactor's high temperature and pressure conditions prevent direct observation of internal parameters, making it necessary to find safer and more efficient ways to simulate the reaction process and monitor key variables. Thanks to the advances in computational technology, numerical simulation has become a widely acknowledged, efficient, and flexible research tool in this field. As a result, many researchers have used this method to study CO_2 methanation [21–24]. However, most existing studies simplify the heat transfer process. They usually only model the tube side reactant flow and ignore the shell side cooling medium. Heat transfer is then calculated based on a constant effective heat transfer coefficient and an isothermal cooling agent temperature. Although this simplification is computationally efficient, it limits the analysis. The constant coefficient cannot reflect the changing temperature distributions on both the tube and shell sides. Since CO_2 methanation takes place in a high pressure, hydrogen rich environment, accurately understanding the reactor tube temperature profile is a critical safety requirement. This is essential to prevent dangerous events like hydrogen explosions caused by local overheating. Thus, fully coupling thermal hydraulic phenomena is vital for precise CO_2 methanation modeling.

CO_2 methanation exhibits rapid reaction kinetics, yet its diffusion within catalysts is constrained. Such limitations manifest as inadequate heat transfer between catalyst particles, which can readily lead to thermal runaway and the formation of hot spots. In numerical simulations of fixed-bed reactors, pseudo-homogeneous or heterogeneous models are widely employed. Schlereth et al. [25] utilized one- and two-dimensional pseudo-homogeneous models for simulations, demonstrating that diffusion and heat conduction within the catalyst bed are critical factors affecting reaction efficiency and temperature distribution.

However, these approaches cannot account for localized flow effects that significantly influence fluid dynamics and heat-mass transfer. Particle-resolved simulations can provide detailed information on velocity and temperature fields. Meanwhile, the random distribution of catalyst particles in the bed complicates fluid flow and heat transfer. Cundall et al. [26] pioneered the Discrete Element Method (DEM), initially developed to study the dynamic behavior of discontinuous media such as rocks. The DEM offers higher resolution and accuracy in simulations. Furthermore, most studies adopt a single-temperature model, assuming thermal equilibrium within the catalyst bed. This approach neglects the effective thermal conductivity between the fluid and catalyst particles. Zhang et al. [27] applied a two-temperature model that considers conjugate heat transfer across multiple regions, including heat transfer between the reactant gas and reactor tube, between the reactor tube and coolant, and between the reactant gas and catalyst bed. It is confirmed that the solver captures the thermal hotspots formed in the reactor. Therefore, in the simulation of CO_2 methanation, obtaining detailed distributions of the flow and temperature fields is crucial for understanding the causes of hotspot formation and for suppressing their development. This understanding provides a basis for the further optimization of reactor design.

In this chapter, we have presented several research findings on CO_2 methanation in shell-and-tube reactors, monolithic reactors, and radial flow reactors, which offers valuable insights into temperature management in fixed bed reactors for CO_2 methanation. These studies adopted a triangular reaction mechanism, in which the reaction kinetics were determined over a Ni/ZrO_2 catalyst.

2. CO_2 methanation in a single tube shell-and-tube reactor

In this section, numerical models of a shell-and-tube reactor were established to systematically explore temperature distributions and reaction performance during the CO_2 methanation process.

2.1 Geometry of the reactor

CO_2 methanation is a complex multiphysics process that integrates fluid dynamics, coupled heat and mass transfer, and heterogeneous catalysis. The shell and tube reactor used here, which has cylindrical symmetry as **Figure 1** schematically shows. Some researches use computationally efficient 2D axisymmetric simulation model [28, 29]. However, here the full 3D model was established. The reactor has three main parts: the interior reaction zone, the cooling shell, and the tube wall. In the interior, a central catalyst bed layer, where methanation happens, is between upper and lower alumina spheres layers. The reactant feed gas goes into the tube side, and the coolant flows through the shell side. Specifically, the coolant is arranged in a counter current configuration relative to the process gas stream.

2.2 Governing equations

The catalyst bed was treated as an isotropic porous medium. To account for the local thermal non-equilibrium between the solid catalyst and fluid phases, a two-temperature model was formulated to describe thermal behavior in this domain [30, 31]. The governing equations were numerically solved using the Finite Volume Method (FVM). The gas phase transport phenomena are governed by conservation equations for mass continuity, momentum, energy, and chemical species, which can be expressed as follows [32]:

$$\rho_f \nabla \bullet \boldsymbol{u} = 0 \tag{1}$$

$$\frac{1}{\varepsilon}\rho_f(\boldsymbol{u} \bullet \nabla)\boldsymbol{u}\frac{1}{\varepsilon} = -\nabla p + \nabla \bullet \boldsymbol{\sigma} + \rho g + S_u \tag{2}$$

$$\rho_f C_{p,f} \boldsymbol{u} \bullet \nabla T_f = \nabla \bullet \left(\kappa_f \nabla T_f\right) + Q_f + S_f \tag{3}$$

$$\rho_f \nabla \bullet (\boldsymbol{u} C_i) = \rho_f \nabla \bullet (D_i \nabla C_i) + J_i \tag{4}$$

where ε, ρ_f, $\boldsymbol{u}$, p, and $\boldsymbol{\sigma}$ represent the porosity, density of the gas phase, velocity, pressure, stress tensor, respectively. C_i, $C_{p,f}$, T_f, and κ_f indicate the concentration of

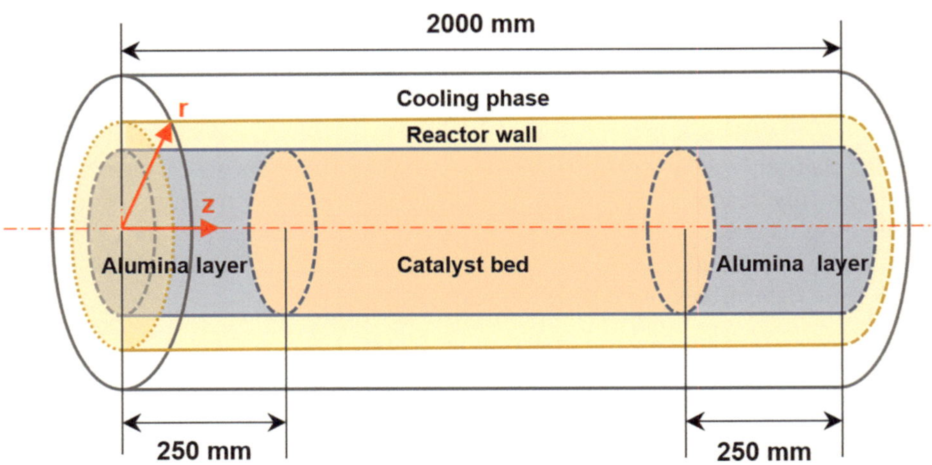

Figure 1.
Axisymmetric geometry of the shell-and-tube reactor simulated.

species i, heat capacity of the gas phase, temperature of the gas phase, thermal conductivity of gas phase, respectively. g is the gravitational acceleration, and D_i stands for the diffusion coefficient of species i. S_u and S_f are the source terms in momentum and temperature equations. J_i and Q_f are the reaction flux of species i and reaction heat, respectively. Under the local thermal non-equilibrium condition, the energy conservation equation governing the solid-phase temperature must be included:

$$\nabla \bullet (\kappa_p \nabla T_p) + S_P = 0 \tag{5}$$

where κ_p is the thermal conductivity, and T_p is the temperature of catalyst pellet. S_p is the source term. S_u In Eq. (2) stands for the resistance caused by the porous catalyst bed, which can be calculated by the Darcy-Forchheimer law:

$$S_u = -\mu_f D\boldsymbol{u} - \frac{1}{2}\rho_f F|\boldsymbol{u}|\boldsymbol{u} \tag{6}$$

$$D = \frac{150}{d_p^2}\frac{(1-\varepsilon)^2}{\varepsilon^3} \tag{7}$$

$$F = \frac{3.5}{d_p}\frac{(1-\varepsilon)}{\varepsilon^3} \tag{8}$$

where D and F are the resistances caused by the viscous and inertial effects, respectively [33, 34]. μ_f is the dynamic viscosity of the fluid phase. d_p is the representative diameter of catalyst pellet. Under the local thermal non-equilibrium condition, the energy source terms S_f and S_p in Eqs. (3) and (5) are written as:

$$S_f = \frac{q_{fp}}{1-\varepsilon}(T_p - T_f) \tag{9}$$

$$S_p = \frac{q_{fp}}{\varepsilon}(T_f - T_p) \tag{10}$$

where q_{fp} is the convective heat transfer coefficient, which can be estimated by Nusselt number (Nu). It should be mentioned that Nusselt number correlation $Nu = 2 + 1.1Re^{0.6}Pr^{1/3}$ proposed by Wakao et al. [35] has been widely used for packed bed. Given the catalyst bed porosity of $\varepsilon = 0.481$ in this study, a modified Nusselt correlation was implemented to ensure simulation fidelity. This formulation maintains accuracy across extended porosity ranges characteristic of packed-bed reactors [36].

$$q_{fp} = \frac{\kappa_f Nu}{d_p^2} \tag{11}$$

$$Nu = \left(2 + \frac{12(1-\varepsilon)}{\varepsilon}\right) + \frac{1}{2}(1-\varepsilon)^{0.5} Re^{0.6} Pr^{\frac{1}{3}} \tag{12}$$

$$Re = \frac{\rho_f u_D d_p}{\mu_f} \tag{13}$$

$$Pr = \frac{C_{p,mix}\mu_{f,mix}}{\kappa_{f,mix}} \tag{14}$$

here, Re and Pr denote the Reynolds and Prandtl numbers, respectively, while $C_{p,mix}$, $\mu_{f,mix}$, and $\kappa_{f,mix}$ represent the heat capacity, dynamic viscosity, and thermal conductivity of the gas mixture.

2.3 Methanation kinetic model

Xu and Froment [37] established a fundamental kinetic formulation incorporating a triangular reaction mechanism that integrates CO_2 methanation, CO methanation, and the reverse water-gas shift (RWGS) reaction. This microkinetic description remains prevalent for modeling both CO_2 methanation and steam methane reforming processes [38, 39]. The reactions are:

CO_2 methanation:

$$CO_2 + 4H_2 \rightleftharpoons CH_4 + 2H_2O \quad \Delta H_r^{298K} = -165\frac{kJ}{mol^{-1}} \quad r_{CO_2Meth} = \frac{\frac{k_{CO_2Meth}}{p_{H_2}^{3.5}}\left(p_{H_2}^4 p_{CO_2} - \frac{p_{CH_4} p_{H_2O}^4}{k_{eq,CO_2Meth}}\right)}{\left(K_{CO}p_{CO} + K_{H_2}p_{H_2} + K_{CH_4}p_{CH_4} + \frac{K_{H_2O}p_{H_2O}}{p_{H_2}}\right)^2} \tag{15}$$

RWGS reaction:

$$CO_2 + H_2 \rightleftharpoons CO + H_2O \quad \Delta H_r^{298K} = 41\, kJ/mol^{-1} \quad r_{RWGS} = \frac{\frac{k_{RWGS}}{p_{H_2}}\left(p_{H_2} p_{CO_2} - \frac{p_{CO} p_{H_2O}}{k_{eq,RWGS}}\right)}{\left(K_{CO}p_{CO} + K_{H_2}p_{H_2} + K_{CH_4}p_{CH_4} + \frac{K_{H_2O}p_{H_2O}}{p_{H_2}}\right)^2} \tag{16}$$

CO methanation:

$$CO + 3H_2 \rightleftharpoons CH_4 + 2H_2O \quad \Delta H_r^{298k} = -206\, kJ/mol^{-1} \quad r_{COMeth} = \frac{\frac{k_{COMeth}}{p_{H_2}^{2.5}}\left(p_{H_2}^3 p_{CO} - \frac{p_{CH_4} p_{H_2O}}{k_{eq,COMeth}}\right)}{\left(K_{CO}p_{CO} + K_{H_2}p_{H_2} + K_{CH_4}p_{CH_4} + \frac{K_{H_2O}p_{H_2O}}{p_{H_2}}\right)^2} \tag{17}$$

within these kinetic expressions, p_i denotes the partial pressure of species, k_i represents the forward rate constant for reaction i, while K_i and K_{eq} correspond to adsorption equilibrium constants and reaction equilibrium constants, respectively. Crucially, as established by Xu and Froment [37] and parameterized for Ni/ZrO_2 catalysts by Choi et al. [40] at 0.6 MPa, these kinetic terms follow an Arrhenius formalism; requisite pre-exponential factors and activation energies are listed in **Table 1**. Importantly, the foundational kinetics were derived for powdered catalysts, necessitating geometric scaling of the intrinsic reaction rate $R_{geo,i}$ to accurately represent pellet-scale behavior within the packed-bed numerical model:

$$R_{geo,i} = \beta \bullet r_i \tag{18}$$

where the scaling factor β quantifies the impact of intraparticle diffusion limitations on the intrinsic reaction kinetics within a catalyst pellet, functioning as a volume averaged metric tied to the catalyst's structural features. Importantly, any changes to

Parameter	Value	Unit
$k_{CO2Meth}$	1.51×10^1	$mol \cdot s^{-1} \cdot g\text{-}cat^{-1} \cdot Mpa^{-1.5}$
$k_{RWGS,0}$	4.18×10^2	$mol \cdot s^{-1} \cdot g\text{-}cat^{-1} \cdot Mpa^{-1}$
k_{COMeth}	4.29×10^3	$mol \cdot s^{-1} \cdot g\text{-}cat^{-1} \cdot Mpa^{-1.5}$
$E_{a,CO2Meth}$	17.80	$kJ \cdot mol^{-1}$
$E_{a,RWGS}$	43.30	$kJ \cdot mol^{-1}$
$E_{a,COMeth}$	50.90	$kJ \cdot mol^{-1}$
$K_{CO,0}$	3.07×10^1	Mpa^{-1}
K_{H2O}	7.35×10^1	Mpa^{-1}
$K_{CH4,0}$	1.04	Mpa^{-1}
$K_{H2O,0}$	1.56×10^3	(−)
ΔH_{CO}	19.40	$kJ \cdot mol^{-1}$
ΔH_{H2}	17.10	$kJ \cdot mol^{-1}$
ΔH_{CH4}	−11.80	$kJ \cdot mol^{-1}$
ΔH_{H2O}	41.20	$kJ \cdot mol^{-1}$

Table 1.
Values of estimated kinetic parameters for Ni/ZrO_2 catalyst under 0.6 MPa [40].

the catalyst's pore structures require a reevaluation of β. This ensures an accurate representation of mass transport constraints under the new conditions. Such precise parameterization is vital for the model to accurately capture the diffusion-limited phenomena that govern reaction rates at the pellet scale [41–43].

In this study, the influence of water in the products of CO_2 methanation is neglected. It is reasonable since the numerical model is developed for high-temperature reaction systems. Under these conditions, water exists as the vapor phase, and its competitive adsorption effects on catalyst active sites are significantly diminished. No water film will form over the catalyst. Concurrently, thermodynamic analysis demonstrates that, within the temperature and pressure ranges employed in the simulations, water's contribution to the equilibrium shift of side reactions, such as the RWGS reaction, is not that pronounced [44].

2.4 Sensitivities of operation parameters

In order to perform the simulations, it is also necessary to determine some boundary conditions. Here, velocity inlets were specified for both the gas and coolant phases at the boundaries, with pressure outlets set for the outflow boundaries. Temperature coupled interfaces were established at three key locations: (i) between the gas phase and catalyst pellets/monolith walls; (ii) between the gas phase and reactor tube inner wall surface; (iii) between the coolant and reactor tube outer wall surface. For porous pellets, the particle diameter was computationally reduced by 1% to improve mesh quality and convergence, a method referenced in Yang et al. [45] and Nijemeisland et al. [46] .

First the impact of different catalyst properties on heat transfer efficiency and reaction performance within the shell-and-tube reactor was investigated [44, 47].

The porous medium was assumed to be homogeneous and isotropic in the model. The effects of different thermal conductivity, porosity, and size of catalyst particles on the gas temperature along the axial bed was obtained. **Figure 2** illustrates the significant impact of the catalyst's heat conductivity. Due to more heat transfer to the reactor wall via the catalyst bed, the catalyst bed's temperature dropped, which in turn lowered the gas phase temperature.

In **Figure 3**, higher porosity led to a higher gas temperature after the peak but slightly decreased the peak value. For instance, at ε_{por} = 0.584, the peak value reduced because the catalyst amount was less than at ε_{por} = 0.481 (higher porosity means fewer catalysts). However, the lower catalyst amount decreased heat transfer to the reactor wall, resulting in a temperature rise after the peak.

As shown in **Figure 4**, the gas temperature variations were almost the same while the catalyst diameter changes. This suggests that the catalyst size is not a key factor in the reactor's temperature distribution. It mainly affects the heat exchange rate

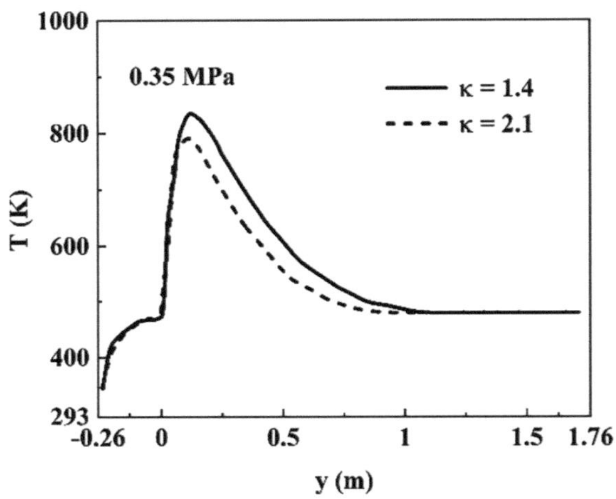

Figure 2.
Effect of thermal conductivity on the temperature of the gas phase in the catalyst bed.

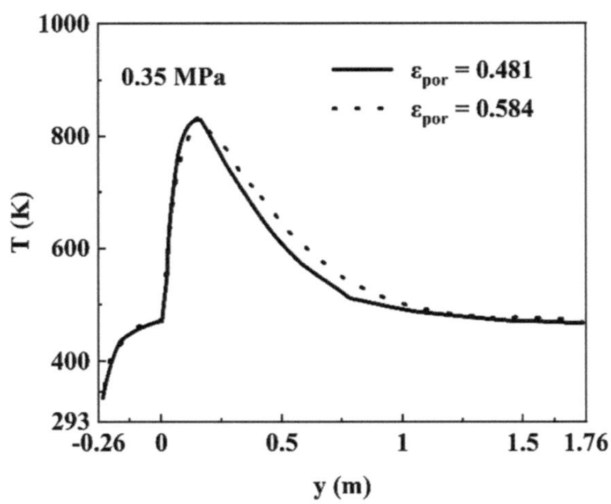

Figure 3.
Effect of porosity on the temperature of the gas phase in the catalyst bed.

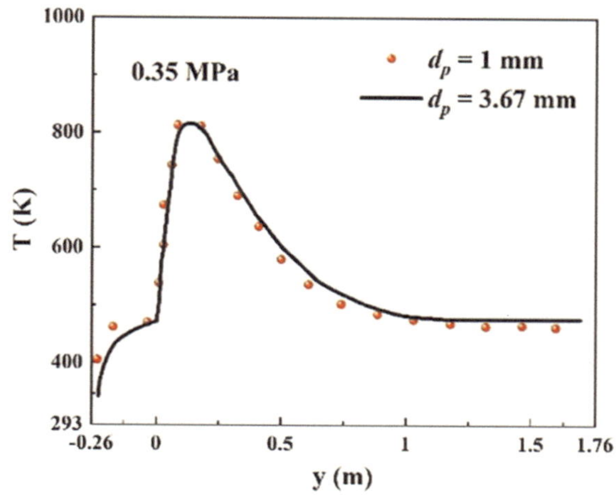

Figure 4.
Effect of catalyst diameter on the temperature of the gas phase in the catalyst bed.

between the gas phase and the catalyst bed, with no relation to wall heat transfer. Therefore, changes in the catalyst size did not affect the heat flux transferred to the reactor wall through the gas and catalyst bed.

2.5 Catalyst dilution

As discussed above, the thermal conductivity of the catalyst significantly impacts tube side heat transfer. Therefore, a catalyst dilution strategy was proposed [48], which involved adding high thermal conductivity inert particles to pure catalyst particles, to enhance heat transfer [49]. As shown in **Figure 5**, the effects of pure catalysts, layered mixing, and random mixing on the reactor temperature distribution were investigated. For the layered mixing, 9 layers and 5 layers were considered.

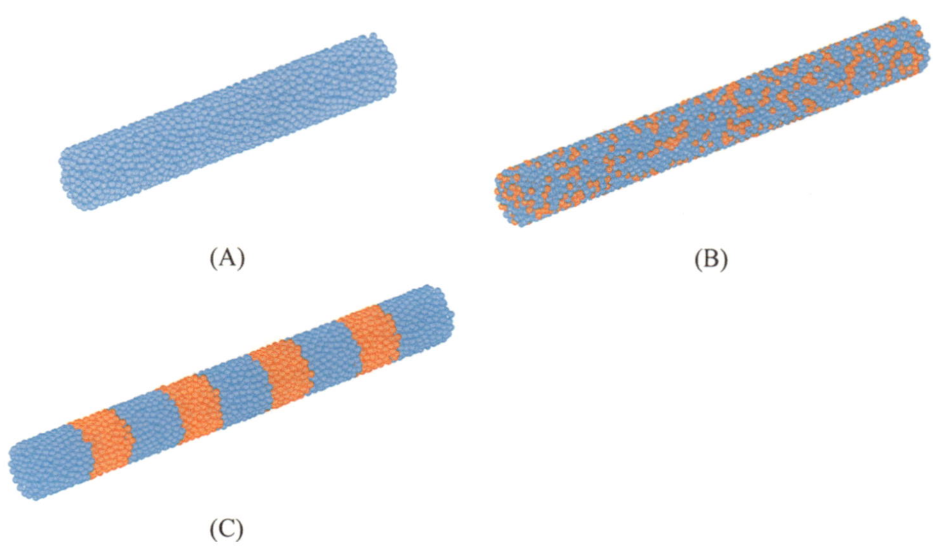

Figure 5.
Packing structure considered: (A) without inert particles; (B) uniformly mixed with inert particle; (C) Inert particle and catalyst in 9 layers. Figure excerpted from literature [48] with permission.

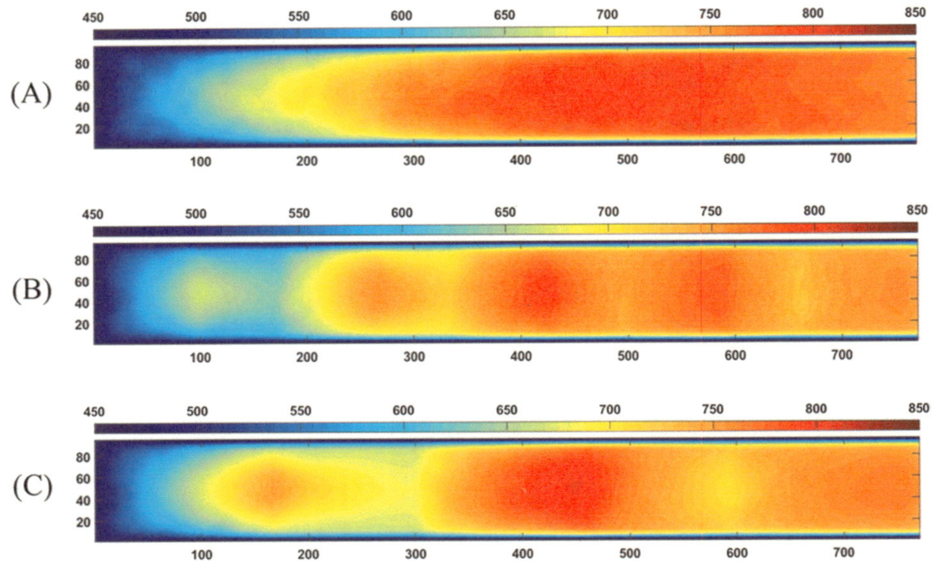

Figure 6.
The temperature distribution in the central cross-section for: (A) Catalyst randomly mixed with inert particle; (B) Inert particle and catalyst in 9 layers; (C) Inert particle and catalyst in 5 layers. Figure excerpted from literature [48] with permission.

All the pellets have a diameter of 3.4 mm. The number of catalyst particles was maintained identical between different structural layers within each layer.

Temperature distributions illustrated in **Figure 6** show that the inert particle dilution technique effectively maintains bed temperatures below 800 K for both mixing and layered configurations. Under the layered dilution approach, temperature contours demonstrate that inert particle stratification creates axial temperature variations in the central region. These variations arise from inconsistent heat transfer across catalyst layers [48], directly reducing the axial thermal uniformity of the reaction bed.

2.6 Reactor with phase change coolant

This part compares the cooling efficiencies of conventional oil coolants and phase change coolants, specifically water. The Volume of Fluid (VOF) method was adopted to simulate the nucleate boiling regime of coolant water [47, 50, 51], aiming to efficiently utilize the latent heat of water evaporation.

Figure 7 illustrates the relationship between peak reactor temperatures and coolant parameters including flow rate and type. For the thermal oil coolant, decreasing the flow rate to 0.5 and 0.1 m^3/h raised peak temperatures to 840 and 980 K, respectively, highlighting the limitations of flow dependent convective cooling [47]. Notably, the saturated water coolant achieved similar peak temperature suppression despite different flow regimes.

Figure 8 further reveals invariant axial temperature profiles along the centerline under varying saturated water flow rates. This confirms fundamentally distinct heat removal mechanics, which are dominated by latent heat exchange during water evaporation rather than sensible heat transfer [47], since the temperatures are almost the same for different water flow rates.

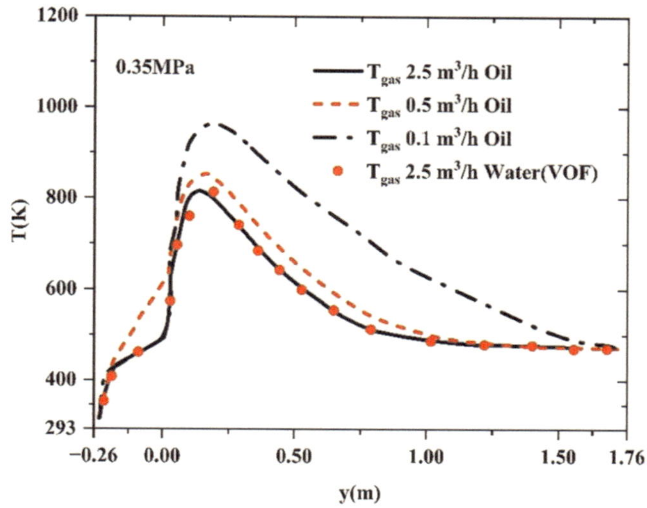

Figure 7.
Gas temperature for different oil flow rates on the shell side.

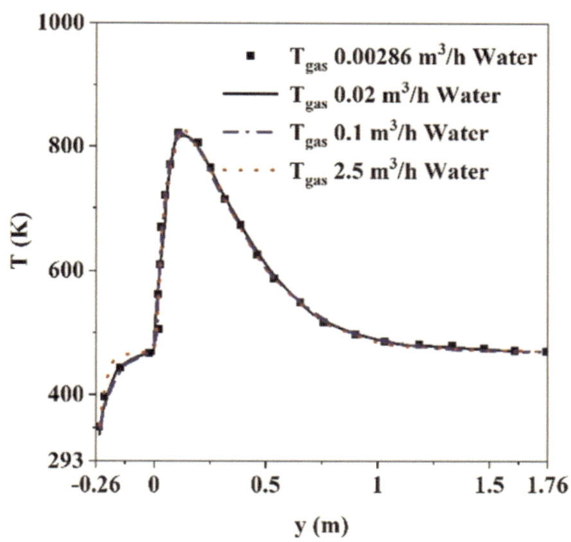

Figure 8.
Gas temperature for different coolant water flow rates on the shell side.

3. CO_2 methanation in monolithic reactor

Compared to randomly packed catalyst bed, structured catalysts exhibit significantly lower pressure drop and enhanced heat and mass transfer characteristics [52]. These traits allow for precise temperature profile regulation in fixed bed reactors. Monolithic catalysts exemplify such structured systems, showing superior reaction performance in various processes and hold particular promise for CO_2 methanation applications.

3.1 Geometry of the monolith

Four different channel dimensions for monolithic catalysts were evaluated, as detailed in **Table 2**. **Figure 9** illustrates the geometry of the monolith reactor simulated.

3.2 Results and discussion

Figure 10 shows the temperature distributions in the central cut plane of monoliths with different channel sizes, indicating that the thermal characteristics depend significantly on geometric parameters. Monoliths with larger channels, which means lower channel density, have more uniform thermal profiles and lower peak temperatures [52]. In contrast, reducing the channel size leads to more significant hotspot formation and higher maximum temperatures. As shown in the **Figure 10**, the radial heat transfer capability within the reactor gradually diminishes as the channel size decreases. Since CO_2 methanation primarily occurs on the surfaces of the monolith channels, a parallel-channel structure can hinder radial heat dissipation. This geometric constraint results in substantial radial thermal gradients, causing significant temperature differences across the cross-section.

The results also indicate that monolithic catalysts achieve higher carbon conversion per unit pressure drop compared to catalyst packed bed. Reducing the channel size of a monolithic catalyst increases the reactive surface area, thereby enhancing carbon conversion. However, this expansion of the reaction area leads to significantly higher reaction intensity, triggering a substantial rise in hotspot temperature. Flow within the channel can be treated as one-dimensional, exhibiting poor radial heat transfer effectiveness. Consequently, heat generated along the central axis is hard to dissipate from the monolith center to the surrounding coolant. Zhang et al. [52]

Case	XL-monolith	L-monolith	M-monolith	S-monolith
Channel size [mm]	8	6	4	2
Wall thickness [mm]	0.5	0.5	0.5	0.5

Table 2.
The monolith with four channel sizes.

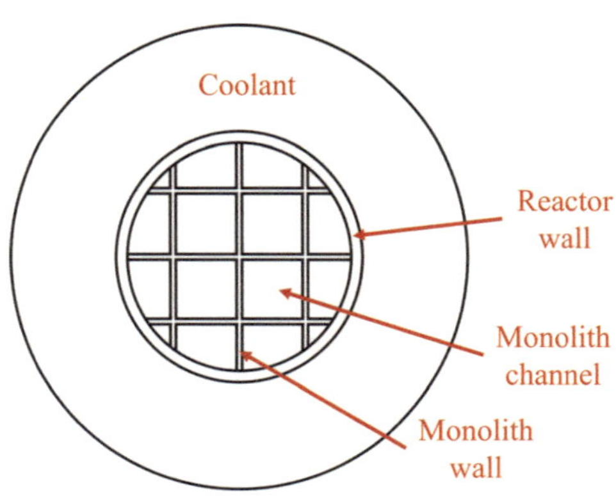

Figure 9.
The geometry of the monolith reactor simulated. The channel size varies in terms of ***Table 2.***

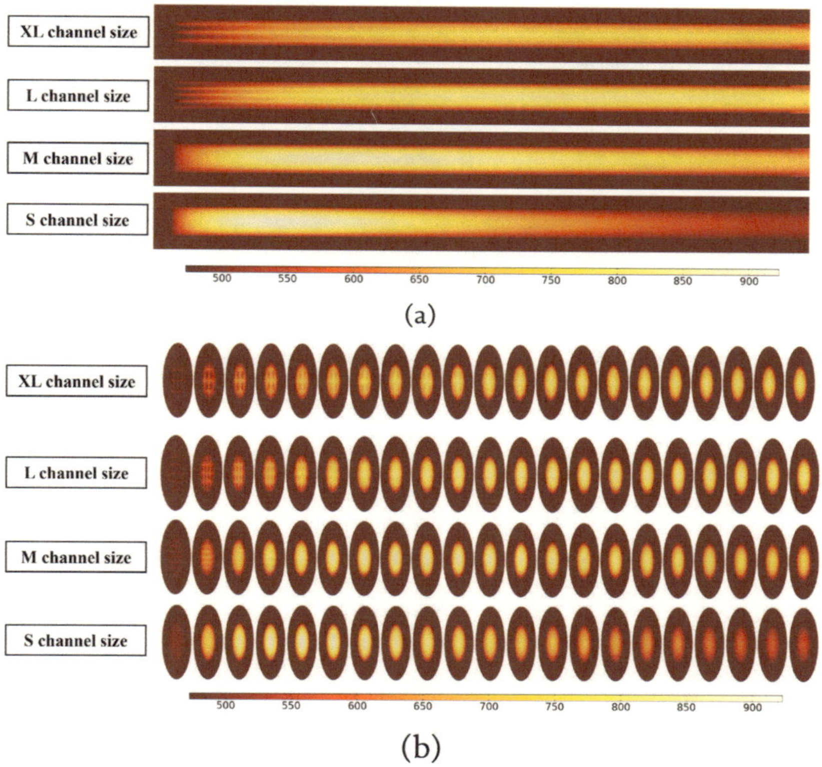

Figure 10.
The temperature distribution for different monolithic reactors: (a) Temperature distribution on the central sectional view; (b) Temperature on the cross view.

proposed a "CHESS" monolith design. The "CHESS" monolith features alternating seals applied to specific channels at the inlet, outlet, and within the monolith channels, creating a chessboard-like pattern. To promote fluid flow mixing between adjacent channels, periodic openings are designed within the channel walls. Sealing blocks are placed to enhance radial heat transfer. Although this structure is more complex than traditional straight-channel monoliths, it can be manufactured using 3D printing. As illustrated in **Figure 11**, the numerical model for the "CHESS" monolith is based on S-monolith model which has the smallest channel size. Since small-scale features like openings and blocks need a finer mesh for resolution, which can easily exceed calculational resource limits. Therefore, only 1/4 of the "CHESS" monolith structure is modeled in the simulation.

As shown in **Figure 12**, the "CHESS" monolith's temperature distribution and flow pattern are presented. The blocked channels considerably affect the monolith's fluid flow. These obstacles impede the flow by redirecting the gas into different channels, thereby creating a flow pattern similar to that on the shell side of a shell-and-tube heat exchanger. Channels next to the blocks have near zero velocity, indicating underutilization as they are sealed by the blocks. The "CHESS" monolith's pressure drop rises significantly to 350 Pa, higher than the S-monolith's and closer to that of packed bed reactors. However, the temperature has been largely reduced compared to that of S-monolith, while maintaining a similar level of carbon conversion [52]. This indicates a better performance behavior of the proposed "CHESS" monolith structure.

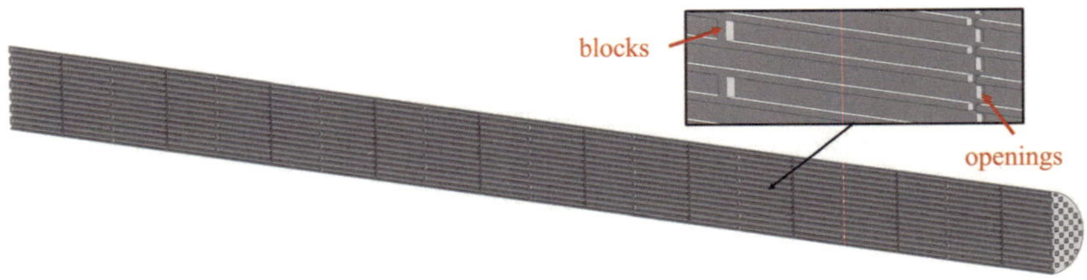

Figure 11.
The structure of the "CHESS" monolith.

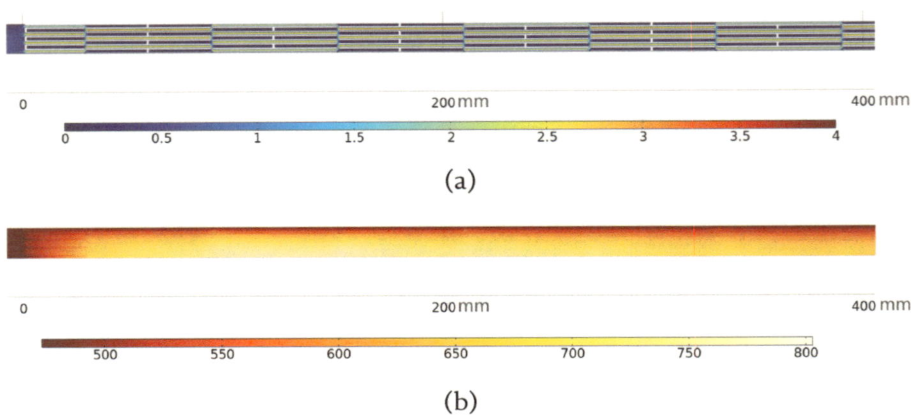

Figure 12.
The flow pattern and temperature distribution of the "CHESS" monolith: (a) Velocity; (b) Temperature distribution.

4. CO_2 methanation in radial flow reactor

4.1 Geometry of the reactor

Four radial flow reactors for CO_2 methanation have been investigated as well [53]. These reactors have different flow configurations shown in **Figure 13**, including centrifugal Z-type, centrifugal Π-type, centripetal Z-type, and centripetal Π-type. They assessed their fluid dynamics, thermal transport, and catalytic performance under the same operating conditions. On the reactor wall, each of the 9 layers features 22 uniformly distributed openings to facilitate gas flow into and out of the catalyst bed. Ni/ZrO_2 particles with a particle size of 3.4 mm was selected as the catalyst and the bed porosity is 0.481.

4.2 Results and discussion

The four reactor configurations show distinct thermal behavior in **Figure 14**. All designs show significant temperature rise in the catalyst bed because of CO_2 methanation's exothermic nature [53]. Notably, the centripetal Z type reactor has unique thermal characteristics, with a clear low temperature area at the base of the catalyst bed. This results from its special heat exchange interface, where incoming low

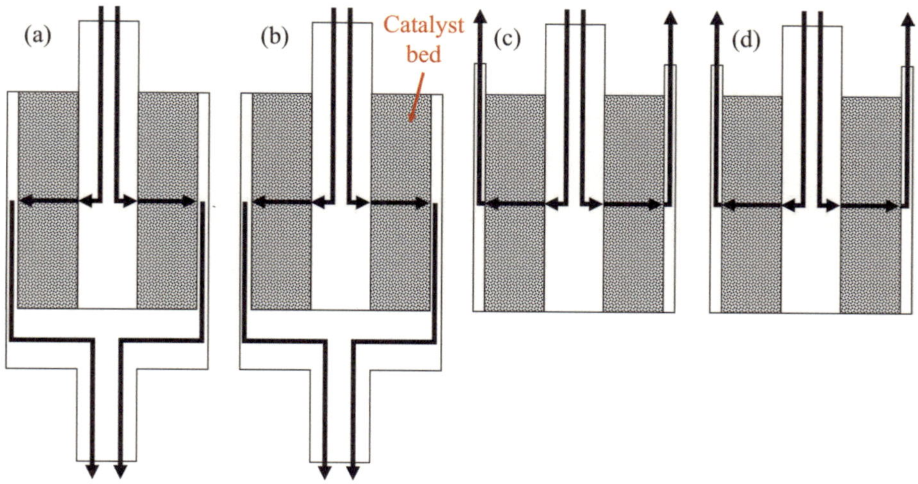

Figure 13.
Four configurations of radial flow fixed bed reactor: (a) Centrifugal Z-type; (b) Centripetal Z-type; (c) Centrifugal Π-type; (d) Centripetal Π-type.

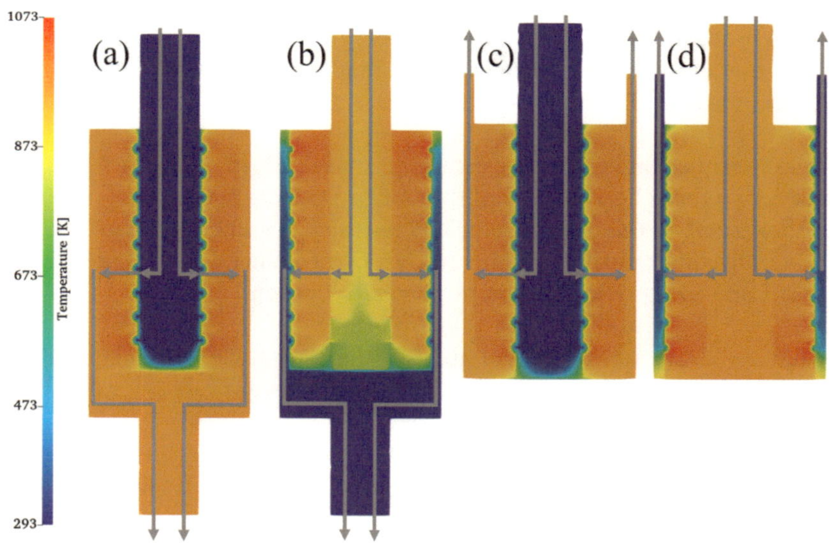

Figure 14.
Temperature distribution in: (a) Centrifugal Z-type; (b) Centripetal Z-type; (c) Centrifugal Π-type; (d) Centripetal Π-type.

temperature feed gas cools the lower catalyst region. Other configurations lack this feature, so they cannot cool the feed gas effectively. In the radial direction, the uniform temperature distribution shows consistent peripheral flow ingress across all designs. In the axial direction, centripetal flow reactors have steeper thermal gradients due to better convective coupling. The feed gas entering through the outer shell has more contact with the catalyst bed, which greatly improves heat transfer efficiency compared to other flow arrangements.

Centripetal configurations, for both Z- and Π- types show abnormal kinetic stratification, with reaction rates violently fluctuating from basal to apical channels. This

	centrifugal Z-type	centripetal Z-type	centrifugal Π-type	centripetal Π-type
CO_2 conversion (X_{CO2})	78.00%	78.39%	78.67%	78.52%
CH_4 selectivity (S_{CH4})	85.22%	88.97%	85.76%	88.65%

Table 3.
Carbon conversion for the radial flow reactors.

thermal sensitivity indicates greater temperature non uniformity, where local hotspots selectively boost CO_2 methanation. Paradoxically, this gradient slightly enhances CH_4 selectivity as shown in **Table 3**. In contrast, centrifugal reactors maintain nearly constant vertical reaction rates, showing superior thermal homogeneity and reaction uniformity despite lower peak selectivity.

5. Conclusions

This chapter introduced the numerical simulations of CO_2 methanation reactors. It systematically analyzed how the catalyst's thermal conductivity, porosity, particle size, and the proposed "CHESS" monolith affect gas temperatures in reactors. It also investigated the temperature distribution and reaction performance of four radial reactor configurations. The main conclusions are summarized as follows:

1. Higher thermal conductivity of catalyst reduces the gas phase temperature by transferring more heat to the reactor wall. It impacts reactor performance more significantly than the pellet size or bed porosity. Adding high thermal conductivity inert particles to the catalyst can enhance heat transfer, with random dilution proving superior to layered dilution.

2. Packed bed could achieve higher absolute carbon conversion, whereas monolithic catalysts demonstrated significantly lower pressure drop. Reduced size of the monolithic channels led to increased pressure loss but improved carbon conversion. Nevertheless, once the channel size declined below a critical value, the conversion achieved per unit pressure drop in the monolith fell beneath the level observed in a pellet packed bed. The newly proposed "CHESS" monolith structure outperformed both the packed bed and the standard straight-channel monolith.

3. In radial flow reactor optimization, Π-type designs show better flow uniformity than Z type ones. Centrifugal flow ensures more uniform velocity distribution than centripetal flow. However, centripetal reactors reduce dead zone volume by 18 25%, improving mass transfer despite slightly more flow maldistribution.

Acknowledgements

This work was partially financially supported by the Science Foundation of China University of Petroleum, Beijing (No.2462021BJRC001), Fuzhou University Qishan Scholar (Overseas Project).

Author details

Wei Zhang[1*], Yixiong Lin[2], Koyo Norinaga[3], Shijie Wu[1], Yuming Zhang[1], Zhewen Chen[1] and Jiazhou Li[1]

1 College of Mechanical and Transportation Engineering, China University of Petroleum-Beijing, Beijing, China

2 College of Chemical Engineering, Fuzhou University, Fuzhou, Fujian, China

3 Department of Chemical Systems Engineering, Graduate School of Engineering, Nagoya University, Furo-cho, Chikusa-ku, Nagoya, Aichi, Japan

*Address all correspondence to: zhang.wei@cup.edu.cn

References

[1] Dell'Amico DB, Calderazzo F, Labella L, et al. Converting carbon dioxide into carbamato derivatives. Chemical Reviews. 2003;**103**:3857-3898

[2] Riduan SN, Zhang Y. Recent developments in carbon dioxide utilization under mild conditions. Dalton Transactions. 2010;**39**:3347-3357

[3] Ma J, Li L, Wang H, et al. Carbon capture and storage: History and the road ahead. Engineering. 2022;**14**:33-43

[4] Hill LB, Li XC, Wei N. CO_2-EOR in China: A comparative review. International Journal of Greenhouse Gas Control. 2020;**103**:103173

[5] Kumar N, Sampaio MA, Ojha K, et al. Fundamental aspects, mechanisms and emerging possibilities of CO_2 miscible flooding in enhanced oil recovery: A review. Fuel. 2022;**330**:125633

[6] Jiang L, Chen Z, Ali SMF, et al. Storing carbon dioxide in deep unmineable coal seams for centuries following underground coal gasification. Journal of Cleaner Production. 2022;**378**: 134565

[7] He Y, Zhu L, Fan J, et al. Life cycle assessment of CO_2 emission reduction potential of carbon capture and utilization for liquid fuel and power cogeneration. Fuel Processing Technology. 2021;**221**:106924

[8] MIT. The Future of Coal. Technology Reports. Massachusetts Institute of Technology; 2007. Available from: https://energy.mit.edu/research/future-coal/

[9] Isah A, Arif M, Hassan A, et al. A systematic review of anhydrite-bearing reservoirs: EOR perspective, CO_2-geo-storage and future research. Fuel. 2022; **320**:123942

[10] Alarcón A, Guilera J, Andreu T. An insight into the heat-management for the CO_2 methanation based on free convection. Fuel Processing Technology. 2021;**213**:106666

[11] Xu Y, Wan H, Du X, et al. Highly active $Ni/CeO_2/SiO_2$ catalyst for low-temperature CO_2 methanation: Synergistic effect of small Ni particles and optimal amount of CeO_2. Fuel Processing Technology. 2022;**236**:107418

[12] Yuan X, Li B, Li B, et al. Performance and stability in CO_2 reforming of methane over neodymium nickel oxide composite catalysts. Fuel Processing Technology. 2021;**211**:106581

[13] Shen W, Zhang Y, Zhao L, et al. Micro-scale simulation and intensification of complex Sabatier reaction system in cylindrical catalyst bed. Fuel. 2021;**287**:119399

[14] Kosaka F, Yamaguchi T, Ando Y, et al. Thermal management of CO_2 methanation with axial staging of active metal concentration in Ni-YSZ tubular catalysts. International Journal of Hydrogen Energy. 2021;**46**:4116-4125

[15] Forzatti P, Lietti L. Catalyst deactivation. Catalysis Today. 1999;**52**: 165-181

[16] Bartholomew CH. Mechanisms of catalyst deactivation. Applied Catalysis A: General. 2001;**212**:17-60

[17] Feng F, Song G, Xiao J, et al. Carbon deposition on Ni-based catalyst with TiO_2 as additive during the syngas methanation process in a fluidized bed reactor. Fuel. 2019;**235**:85-91

[18] Esa YAM, Sapawe N. A short review on carbon dioxide (CO_2) methanation

process. Materials Today: Proceedings. 2020;**31**:394-397

[19] Kopyscinski J, Schildhauer TJ, Biollaz SMA. Methanation in a fluidized bed reactor with high initial CO partial pressure: Part I—Experimental investigation of hydrodynamics, mass transfer effects, and carbon deposition. Chemical Engineering Science. 2011;**66**: 924-934

[20] Li J, Zhou L, Li P, et al. Enhanced fluidized bed methanation over a Ni/ Al_2O_3 catalyst for production of synthetic natural gas. Chemical Engineering Journal. 2013;**219**:183-189

[21] Alarcón A, Guilera J, Andreu T. CO_2 conversion to synthetic natural gas: Reactor design over Ni–Ce/Al_2O_3 catalyst. Chemical Engineering Research and Design. 2018;**140**:155-165

[22] Chein RY, Chen WY, Yu CT. Numerical simulation of carbon dioxide methanation reaction for synthetic natural gas production in fixed-bed reactors. Journal of Natural Gas Science and Engineering. 2016;**29**: 243-251

[23] Kiewidt L, Thöming J. Predicting optimal temperature profiles in single-stage fixed-bed reactors for CO_2-methanation. Chemical Engineering Science. 2015;**132**:59-71

[24] Ducamp J, Bengaouer A, Baurens P. Modelling and experimental validation of a CO_2 methanation annular cooled fixed-bed reactor exchanger. The Canadian Journal of Chemical Engineering. 2017;**95**:241-252

[25] Schlereth D, Hinrichsen O. A fixed-bed reactor modeling study on the methanation of CO_2. Chemical Engineering Research and Design. 2014; **92**:702-712

[26] Cundall PA. A computer model for simulating progressive, large-scale movements in blocky rock systems. In: ISRM International Symposium. ISRM; 1971

[27] Zhang W, Machida H, Takano H, et al. Computational fluid dynamics simulation of CO_2 methanation in a shell-and-tube reactor with multi-region conjugate heat transfer. Chemical Engineering Science. 2020;**211**:115276

[28] Peláez R, Marín P, Díez FV, et al. Direct synthesis of dimethyl ether in multi-tubular fixed-bed reactors: 2D multi-scale modelling and optimum design. Fuel Processing Technology. 2018;**174**:149-157

[29] Chibani A, Merouani S, Bougriou C, et al. Heat and mass transfer during the storage of hydrogen in LaNi5-based metal hydride: 2D simulation results for a large scale, multi-pipes fixed-bed reactor. International Journal of Heat and Mass Transfer. 2020;**147**:118939

[30] Hsu CT. A closure model for transient heat conduction in porous media. Journal of Heat Transfer. 1999; **121**:733-739

[31] Gao X, Zhu YP, Luo Z. CFD modeling of gas flow in porous medium and catalytic coupling reaction from carbon monoxide to diethyl oxalate in fixed-bed reactors. Chemical Engineering Science. 2011;**66**: 6028-6038

[32] Lin Y. Numerical study on chemical reactions and transport phenomenon in the solid catalyst fixed bed for methanation of carbon dioxide and dry reforming of methane [diss.]. Nagoya, Japan: Nagoya University;

[33] Ergun S, Orning AA. Fluid flow through randomly packed columns and

fluidized beds. Industrial & Engineering Chemistry. 1949;**41**:1179-1184

[34] Ergun S. Fluid flow through packed columns. Chemical Engineering Progress. 1952;**48**:89-94

[35] Wakao N, Kaguei S, Funazkri T. Effect of fluid dispersion coefficients on particle-to-fluid heat transfer coefficients in packed beds: Correlation of Nusselt numbers. Chemical Engineering Science. 1979;**34**:325-336

[36] Kuwahara F, Shirota M, Nakayama A. A numerical study of interfacial convective heat transfer coefficient in two-energy equation model for convection in porous media. International Journal of Heat and Mass Transfer. 2001;**44**:1153-1159

[37] Xu J, Froment GF. Methane steam reforming: II. Diffusional limitations and reactor simulation. AICHE Journal. 1989; **35**:97-103

[38] Ngo SI, Lim YI, Lee D, et al. Flow behaviors, reaction kinetics, and optimal design of fixed-and fluidized-beds for CO_2 methanation. Fuel. 2020;**275**:117886

[39] Sun D, Khan FM, Simakov DSA. Heat removal and catalyst deactivation in a Sabatier reactor for chemical fixation of CO_2: Simulation-based analysis. Chemical Engineering Journal. 2017;**329**:165-177

[40] Choi C, Khuenpetch A, Zhang W, et al. Determination of kinetic parameters for CO_2 methanation (sabatier reaction) over Ni/ZrO_2 at a stoichiometric feed-gas composition under elevated pressure. Energy & Fuels. 2021;**35**:20216-20223

[41] Wang Q, Yang C, Wang H, et al. Optimization of process-specific catalytic packing in catalytic distillation process: A multi-scale strategy. Chemical Engineering Science. 2017;**174**: 472-486

[42] Hafeez S, Aristodemou E, Manos G, et al. Computational fluid dynamics (CFD) and reaction modelling study of bio-oil catalytic hydrodeoxygenation in microreactors. Reaction Chemistry & Engineering. 2020;**5**:1083-1092

[43] Zhao Y, Shi Y, Ye G, et al. Optimization of catalyst pellet structures and operation conditions for CO methanation. Chinese Journal of Chemical Engineering. 2021;**40**:106-113

[44] Lin Y, Zhang W, Machida H, et al. CFD simulation of the Sabatier process in a shell-and-tube reactor under local thermal non-equilibrium conditions: Parameter sensitivity and reaction mechanism analysis. International Journal of Hydrogen Energy. 2022;**47**: 15254-15269

[45] Yang X, Wang S, Zhang K, et al. Evaluation of coke deposition in catalyst particles using particle-resolved CFD model. Chemical Engineering Science. 2021;**229**:116122

[46] Nijemeisland M, Dixon AG. Comparison of CFD simulations to experiment for convective heat transfer in a gas–solid fixed bed. Chemical Engineering Journal. 2001;**82**:231-246

[47] Fukumoto K, Zhang W, Mizoguchi R, et al. CFD simulation of CO_2 methanation through the Sabatier reaction in a shell-and-tube reactor incorporating phase change on the shell side. Fuel. 2023;**349**:128126

[48] Lin Y, Yang C, Choi C, et al. Inhibition of temperature runaway phenomenon in the Sabatier process using bed dilution structure: LBM-DEM simulation. AICHE Journal. 2021;**67**: e17304

[49] Hong GH, Noh YS, Park JI, et al. Effect of catalytic reactor bed dilution on product distribution for Fischer-Tropsch synthesis over Ru/Co/Al_2O_3 catalyst. Catalysis Today. 2018;**303**:136-142

[50] Sato Y, Niceno B. Pool boiling simulation using an interface tracking method: From nucleate boiling to film boiling regime through critical heat flux. International Journal of Heat and Mass Transfer. 2018;**125**:876-890

[51] Nabil M, Rattner AS. interThermalPhaseChangeFoam—A framework for two-phase flow simulations with thermally driven phase change. SoftwareX. 2016;**5**:216-226

[52] Zhang W, Lin Y, Zhang Y, et al. Regulation of temperature distribution in fixed bed reactor for CO_2 methanation through "CHESS" monolith structure catalyst. Applied Thermal Engineering. 2024;**236**:121826

[53] Zhang W, Lin Y, Norinaga K. Insights into structure-performance relationship in radial flow fixed bed reactor for CO_2 methanation. International Journal of Hydrogen Energy. 2023;**48**:24594-24606

Chapter 7

Combustion Characteristics and Flame Stability of Hydrogen-Enriched Natural Gas for Domestic Porous Media Burners

Chen Yang, Jie Niu, Yiyu Chen and Zhongmin Wan

Abstract

Existing domestic gas burner are restricted by dual limitations of gas interchangeability and flashback limits, typically being limited to low HENG. Elevated hydrogen ratios degrade flame stability, forming the critical bottleneck constraining the development of HENG technology. A partially premixed porous media gas stove suitable for high HENG was designed in this study. The effects of material and structural parameters of homogeneous porous media on combustion characteristics under varying hydrogen blending ratios were systematically investigated, while the transition mechanism of combustion modes under high hydrogen blending conditions was theoretically analyzed. The results demonstrate that with increasing hydrogen blending ratios, partial reactions were shifted toward the interior of porous media, inducing significant variations in OH radical concentration that triggered combustion mode transitions and ultimately substantially influenced combustion characteristics. The threshold HENG for combustion mode transition was closely correlated with porous media materials and pore density parameters, where higher pore density enabled accommodation of wider hydrogen blending ranges.

Keywords: domestic gas burner, hydrogen-enriched natural gas, porous diffusion combustion, combustion characteristics, flame stability

1. Introduction

Hydrogen energy plays a pivotal role in achieving the "dual-carbon" goals and advancing the "Three Highs and Four New" strategic development [1]. Hydrogen-enriched natural gas (HENG) combustion technology enables the direct application of hydrogen in various fields, including transportation, power generation, and residential gas utilization, significantly facilitating the development of hydrogen energy [2]. A comparison of the key physical and chemical characteristics between methane and hydrogen is presented in **Table 1**. While the two gases share many similarities, they exhibit notable differences in certain parameters. The addition of hydrogen to natural gas is demonstrated in **Table 1** to increase flame propagation velocity while decreasing the required minimum ignition energy of the resulting fuel blend. Additionally, it broadens the flammability limits of the mixture [3].

Properties	Methane	Hydrogen	Unit
Density	0.6801	0.0852	kg/m^3
Lean flammability limit	0.53	0.14	/
Flammability limits in air	5 ~ 15.98	4 ~ 76.51	vol%
Ignition energy	33	2	10^{-5} J
Lower heating value (LHV)	34.04	10.23	MJ/m^3
Higher heating value (HHV)	37.77	12.09	MJ/m^3
Laminar flame speed	0.4	2.1	m/s
Fuel diffusion coefficient	0.16	0.61	cm^2/s

Table 1.
Property comparison between methane and hydrogen.

The domestic gas industry has experienced significant advancement alongside the growing prevalence of gas burner, making the enhancement of thermal performance coupled with substantial reduction in combustion emissions a primary research priority. Extensive and in-depth studies have been conducted by domestic and international scholars on critical aspects of HENG burner, including flame characteristics, thermal efficiency, and emission performance.

Meziane et al. [4] conducted CFD simulations to examine the effects of hydrogen blending on combustion characteristics in Rich-Quench-Lean (RQL) can combustors under both fuel-rich and fuel-lean conditions. Their results demonstrated a significant reduction in CO and NOx emissions with increasing hydrogen fractions. Sun et al. [5] evaluated the performance of domestic gas burner with 0–20% hydrogen blending, observing shortened flame length, improved thermal efficiency, reduced CO emissions, though with increased NOx production. Experimental studies by Fang et al. [6] on radiant porous media burners revealed that hydrogen blending below 25% shortened flame length and reduced CO emissions without significantly affecting NOx levels. Zhao et al. [7] reported accelerated reaction rates in gas burner with hydrogen addition, leading to shorter ignition times but increased flashback risks at 50% hydrogen fraction, with no notable thermal efficiency improvement. In contrast, Chen et al. [8] observed decreasing thermal efficiency with higher hydrogen content in porous media burners. Choundhury et al. [9] documented reduced NOx emissions in conventional burners with hydrogen addition, while Sorgulu et al. [10] reported substantially increased NO but decreased CO emissions.

In summary, the academic community has acquired preliminary understanding of HENG applications in domestic gas burner. The effects of low HENG on flame morphology, thermal efficiency, and pollutant emissions have been partially established, with consistent observations of flame length reduction and CO emission decrease. However, no consensus has been reached regarding the impacts on thermal efficiency and NOx formation, requiring further experimental investigation. Most existing studies, constrained by gas interchangeability requirements and flashback limits in conventional burner, have focused exclusively on low hydrogen blending ratios (typically below 20%). This approach has significantly restricted the carbon reduction potential of HENG. The development of novel combustion technologies to expand hydrogen blending applicability is urgently needed.

As the predominant third-generation combustion solution succeeding regenerative systems, porous media combustion delivers significant improvements in energy utilization while suppressing pollutant emissions by virtue of its engineered structural design. Pan et al. [11] experimentally examined the performance of microscale porous burners, revealing that porous structures enhance flame stabilization and yield superior energy conversion efficiency. Pantangi et al. [12] investigated combustion characteristics of LPG porous media radiant cooktops, documenting a 3% thermal efficiency improvement along with significantly reduced emissions compared to conventional burners. Trimis et al. [13] examined pollutant emission characteristics of porous media burners, revealing significantly lower carbon emissions with CO concentrations of only 10 ppm and NOx levels between 2 and 20 ppm. Mital et al. [14] analyzed fluid processes in combustion systems, confirming the effectiveness of porous media in waste heat recovery and exhaust temperature reduction. Bouma et al. [15] employed a one-dimensional two-temperature model incorporating radiation effects, showing that porous media structures can substantially reduce both CO and NOx emissions compared to conventional free-flame combustion. Laphirattanakul et al. [16] demonstrated the flame stabilization capability of hollow cylindrical ceramic foams, revealing that the thermal conduction properties of porous media are directly influenced by material type and pore structure. Herrera et al. [17] studied thermal efficiency under lean combustion conditions, reporting 7–14% improvement compared to free-flame burners. In summary, porous media combustion demonstrates significant efficacy in stabilizing flames, improving thermal efficiency, and reducing pollutant emissions in gas burners. This technology is expected to simultaneously expand hydrogen blending applicability while improving flame stability.

A novel porous media gas burner was developed in this work for stable operation with HENG at elevated blending ratios. Through combined experimental investigation and numerical simulation, the influence of porous media materials and structural parameters on combustion characteristics was investigated under different HENG. The transformation mechanism of combustion modes under high hydrogen proportions was theoretically analyzed. Furthermore, Fundamental mechanisms underlying flashback occurrences in gas burner were elucidated, establishing essential theoretical frameworks and technical foundations for deploying high-ratio HENG in domestic gas appliances.

2. Combustion experimental system

2.1 Construction of the test platform

The experimental test platform was constructed as the fundamental basis for combustion research and performance evaluation. In accordance with national standard requirements, a combustion test platform for HENG porous media gas burner was designed and established, as illustrated in **Figure 1(b)**. The combustion system shown in **Figure 1(a)** incorporated a gas flow controller, CCD camera, data acquisition system, composite gas analyzer, porous media burner, gas mixer, gas alarm device, along with natural gas and hydrogen cylinders. Methane and hydrogen were delivered from separate high-pressure cylinders, their flow rates accurately regulated using Alicat flow controllers. Prior to entering the porous media combustion chamber, the two fuel gases underwent homogenization in a high-precision mixing apparatus to achieve uniform compositional distribution.

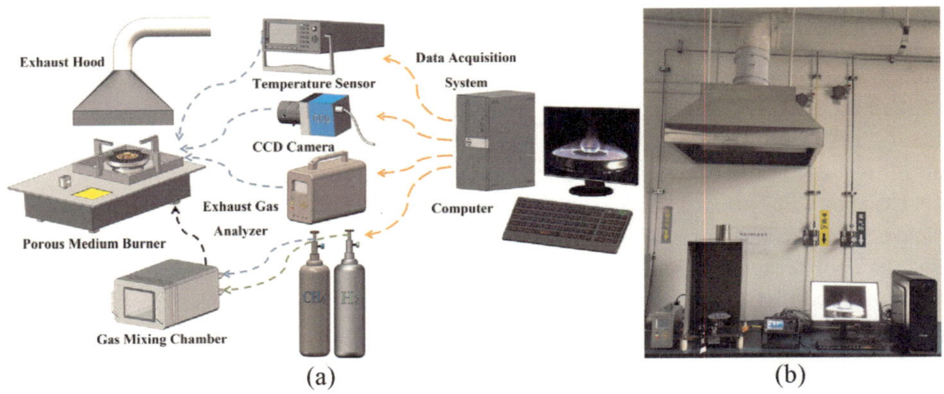

Figure 1.
Schematic of the combustion experimental system. (a) Diagram of the experimental setup (b) Combustion test platform.

2.2 System equipment selection

Key instrumentation parameters for the combustion system are specified in **Table 2**.

The 0–10 SLM range was selected as the optimal flow range based on the methane and hydrogen flow rates presented in **Table 3**.

Measurement apparatus	Type	Accuracy
Thermal Sensor	Type K Thermocouple	0–900°C ± 1.5°C
CCD	MER-132-30UC	1292(H) × 964(V) 30fps
Gas Flow Controller	Alicat	0–10 SLM ± 0.2%
Gas Analyzer	GT2000-LF5	CO: 0–1000 ppm, NOx: 0–100 ppm
Data Acquisition System	DAQ970A + DAQ901A	Maximum Scanning Speed 480 channel/s
		Maximum Sampling Frequency 800 kSa/s
Gas Mixing Apparatus	Ld-005	20–2000 ml/min Uncertainty <1%

Table 2.
Specifications of major experimental devices.

Hydrogen blending ratio (%)	0	5	10	15	20	25	30	35	40	45	50	55	60
Methane Flow Rate (L/min)	6.37	6.27	6.16	6.05	5.92	5.79	5.64	5.48	5.30	5.11	4.89	4.65	4.39
Hydrogen Flow Rate (L/min)	0	0.33	0.68	1.07	1.48	1.92	2.42	2.95	3.54	4.18	4.89	5.69	6.58

Table 3.
Experimental conditions.

3. Combustion characteristics of porous media domestic gas burner

3.1 Flame morphology in HENG combustion

To thoroughly investigate the variation of flame stability with hydrogen blending ratios, an Al_2O_3@60PPI porous medium was analyzed. As illustrated in **Figure 2**, for cases where hydrogen accounted for fewer than 30% of the mixture, the lateral flame fluctuation indicated minimal changes in flame morphology among different groups, demonstrating strong combustion stability. As the proportion of hydrogen in the fuel mixture rises, the flame length was significantly shortened, the internal flame structure became sharper, and the luminosity of the porous medium gradually intensified while the flame color transitioned from pale blue to yellowish tones. Three mechanistic pathways explain these observations. First, intensified radiation from the porous matrix indicates surface-localized exothermic reactions, where thermal energy transfers to the porous structure increases both temperature and radiant intensity. Second, the initial blue emission results from CH* radical chemiluminescence [18]. Third, the observed yellowing suggests incomplete combustion conditions. The transition to yellowish flame resulted from incomplete combustion. Increasing hydrogen content reduced fuel density and consequently decreased primary air flow, creating oxygen-deficient conditions that prevented complete fuel combustion, manifesting as yellowish flames.

Experimental results in **Figure 2** reveal distinct flame morphology alterations when hydrogen concentrations surpass 35%, contrasting markedly with behavior observed at sub-30% mixtures. The porous media region darkens, while the external flame appears dark blue and levitates above the porous medium. A yellow flame

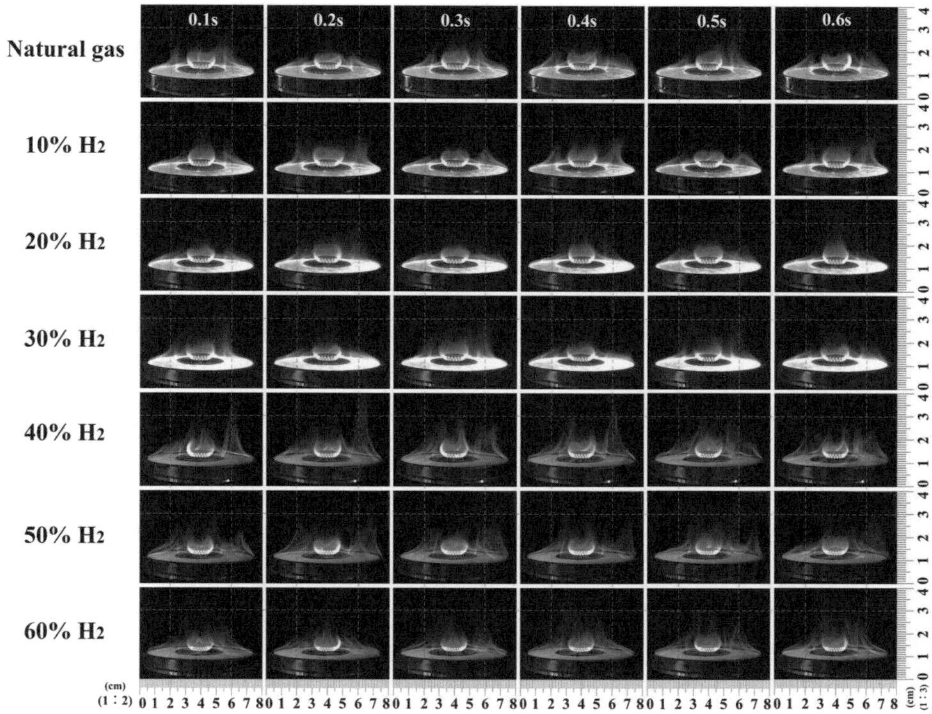

Figure 2.
Flame morphology in the Al_2O_3@60PPI porous media burner.

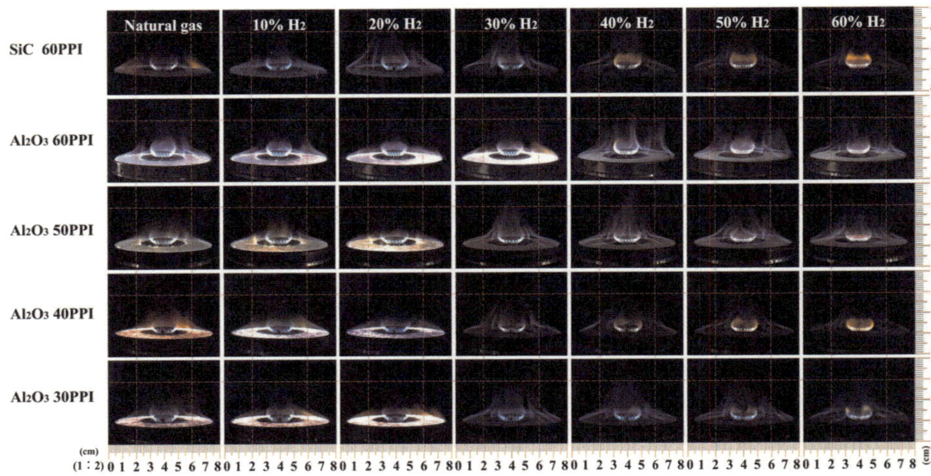

Figure 3.
Flame morphology of Al_2O_3 under different structural parameters.

persists in the central combustion zone, indicating a transition in combustion mode. This phenomenon is attributed to two primary factors. Hydrogen combustion velocity is approximately eight times higher than natural gas. At excessive hydrogen ratios, the markedly increased flame speed prevents flame stabilization within the porous media, causing flashback into the burner interior before being pushed back to the surface, resulting in the darkened porous region. The central combustion zone, representing the inner flame region, suffers from insufficient secondary air supply through the limited bottom air inlets, leading to incomplete combustion and yellow flame formation. When the hydrogen ratio surpasses 40%, further increases cause additional flame shortening, reddening of the inner flame cover, more pronounced yellow flames, while the porous media remains dark and unchanged. These observations confirm that both inner and outer flames gradually lift with increasing hydrogen content, and the oxygen-deficient conditions exacerbate incomplete combustion.

Figure 3 presents the variations in flame morphology with increasing hydrogen blending ratios for gas burner employing SiC and Al_2O_3 porous media materials of different pore densities. As clearly demonstrated, when the hydrogen ratio reaches certain thresholds, all five porous media configurations exhibit significant flame pattern alterations. At specific hydrogen concentrations, the inner burner cap gradually reddens due to reduced flow resistance in low pore density media with larger pore diameters. This accelerated flow velocity decreases mixing time between gas and secondary air, resulting in incomplete combustion that generates carbon monoxide and unburned carbon particles. Thermal radiation from these carbon particles produces yellowish or reddish flames. Al_2O_3 media demonstrate a wider critical hydrogen blending range than SiC at equivalent pore densities. For identical materials, higher pore density porous media enable broader critical hydrogen ranges, enhanced combustion stability, and more complete combustion.

3.2 Pollutant emissions and thermal efficiency in HENG combustion

The flame temperature measurement methodology in this study was conducted by selecting six spatial points along the vertical axis of the burner (x = 0 mm, y = 0 mm; z = 20, 40, 60, 80, 100, and 120 mm) once stable combustion was achieved at each hydrogen blending ratio. Under steady combustion conditions, flame temperatures were acquired at multiple locations using K-type thermocouples for each hydrogen concentration level. Among the six measurement positions, the highest temperature reading was identified as the characteristic peak flame temperature for each specific fuel mixture. The experimental configuration for temperature measurement is presented in **Figure 4**.

Figure 5 demonstrates a consistent positive correlation between hydrogen enrichment levels and peak flame temperature under constant combustion conditions. This phenomenon is attributed to the fewer intermediate reaction steps between H_2 and O_2 compared to the more complex CH_4-O_2 reaction mechanism, resulting in more vigorous oxidation and faster heat release rates for hydrogen combustion [19]. The addition of hydrogen elevates combustion temperatures until a critical 35% blending ratio is reached. At this threshold, the transition in combustion mode induces a sudden temperature decrease, with peak values dropping from 1540 K to 1420 K. Subsequently, with further hydrogen enrichment, the temperature resumes its upward trend through the same mechanism responsible for the initial temperature increase prior to the combustion mode shift.

The thermal efficiency represents the ratio of the chemical energy of fuel converted into effective heat by a gas stove per unit time, reflecting the energy utilization rate of the appliance. A higher thermal efficiency indicates better combustion performance. The specific calculation formula (1) is as follows:

$$\eta = \frac{m \times c \times (t_2 - t_1)}{(V_1 \times Q_1 + V_2 \times Q_2)} \times \frac{273 + t_g}{288} \times \frac{101.3}{p_{amb} - p_m - s} \times 100 \tag{1}$$

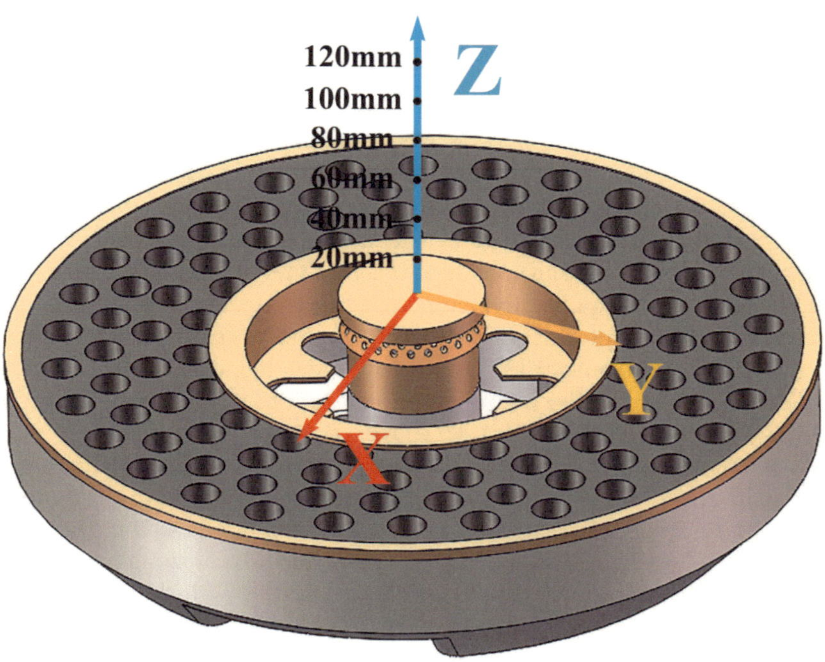

Figure 4.
Schematic diagram of flame temperature monitoring points.

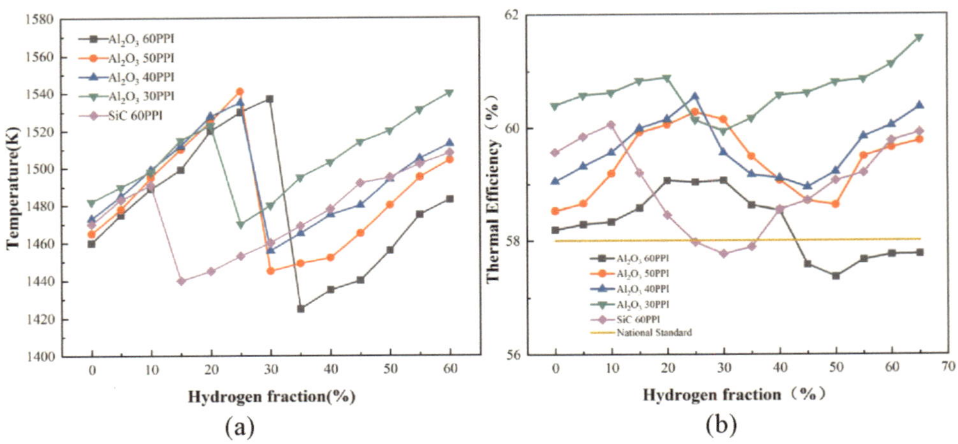

Figure 5.
Variation of temperature and thermal efficiency at different HENG. (a) Temperature (b) Thermal efficiency.

η—The thermal efficiency (%), c—The specific heat capacity of water (MJ/(kg·°C)), t_1—The initial water temperatures (°C), t_2—The final water temperatures (°C), V_1—The methane consumption during combustion (m^3), V_2—The hydrogen consumption during combustion (m^3), Q_1—The lower heating value of methane at standard conditions (15°C, 101.3 kPa), Q_2—The lower heating value of hydrogen at standard conditions (15°C, 101.3 kPa), t_g—The gas temperature inside the flow meter during the experiment (°C), p_{amb} The ambient atmospheric pressure (kPa), p_m—The measured static pressure of gas inside the flow meter (kPa), s—The saturated water vapor pressure at temperature t_g (kPa).

The variation in measured thermal efficiency with structural parameters at different HENG is presented in **Figure 5(b)** for SiC and Al_2O_3 porous media materials. Using the Al_2O_3@60PPI porous media burner as an example, the flame transition point is observed at 35–40% hydrogen blending ratio. The thermal efficiency is found to generally increase with rising hydrogen content when the combustion mode remains unchanged. Following combustion mode transition, an initial decrease in thermal efficiency is observed before gradual recovery. The initial thermal efficiency improvement is attributed to hydrogen enrichment, which enhances H/O/OH radical concentrations in the flame. The enhanced reaction kinetics and accelerated flame front propagation substantially augment convective heat transfer from high-temperature combustion products to the cooking surface, consequently improving thermal efficiency of the burner [3]. At 35% hydrogen blending, thermal efficiency begins to decline. While higher hydrogen concentrations strengthen chemical reactions R25($H + CH_3(+M) \rightleftharpoons CH_4(+M)$), R27($OH + H_2 \rightleftharpoons H + H_2O$) and R14 ($H + O_2 \rightleftharpoons O + H_2$) and flame speed, flame lift-off from the porous media surface is simultaneously caused. The porous media zone demonstrated significant darkening, accompanied by distinct flame attenuation and a marked reduction in radiative heat transfer capacity. Consequently, the heating mechanism transitions from combined conduction-radiation to purely flame-based heat transfer, with the latter dominating system performance and decreasing overall efficiency [20]. With further hydrogen enrichment, continued enhancement of chemical reaction intensity and flame propagation speed is observed under the flame-based heating mechanism, resulting in subsequent thermal efficiency recovery.

As shown in the **Figure 5(b)**, the measured thermal efficiency of the Al_2O_3-based porous media decreases with increasing pore density. When comparing Al_2O_3@60PPI and SiC@60PPI, the latter exhibits higher thermal efficiency due to the greater total emissivity of SiC, which aligns with the findings reported by Muthukumar et al. [21], thereby validating the experimental reliability. Overall, as the hydrogen blending ratio increases, the thermal efficiency of all porous media burners follows a consistent trend: an initial rise, followed by a decline, and then a subsequent increase. Furthermore, before any combustion mode transition occurs, the measured thermal efficiency for each porous media burner exceeds 58%, meeting the requirements specified in the national standard [22].

Figure 6 illustrates the variations in CO emissions from the gas burner with increasing hydrogen blending ratios when using SiC and Al_2O_3 porous media materials, across different pore densities. Experimental investigations on the Al_2O_3@60PPI porous burner demonstrate distinct emission characteristics under hydrogen enrichment conditions. Within the hydrogen blending ratio range of 0–30%, CO emissions demonstrated a decreasing trend while NOx emissions showed an increasing trend with elevated hydrogen addition. A notable transition occurs at 35% hydrogen concentration, where substantial alterations in flame structure develop alongside a dramatic CO concentration jump from 175 to 612 ppm, while NOx production shows an inverse decreasing pattern. Subsequent hydrogen enrichment beyond this critical ratio leads to renewed CO abatement and restored NOx growth. The dual-stage CO mitigation phenomenon primarily originates from two factors [23]: the diminishing carbon fraction in the fuel mixture and hydrogen catalytic enhancement of hydrocarbon oxidation processes, which effectively drives the conversion of carbon monoxide to carbon dioxide according to previous studies. The intermediate CO surge at 35% hydrogen content stems from fundamental changes in combustion characteristics. During this transition, the attenuation of radiative heat transfers due to porous media darkening and compromised thermal exchange performance promotes incomplete oxidation. These effects are compounded by hydrodynamic instability manifested through flame front elongation and irregular propagation patterns, where fuel supply rates surpass the flame propagation velocity [24]. This imbalance creates localized regions with inadequate residence duration and unfavorable thermal conditions, ultimately resulting in increased carbon monoxide formation.

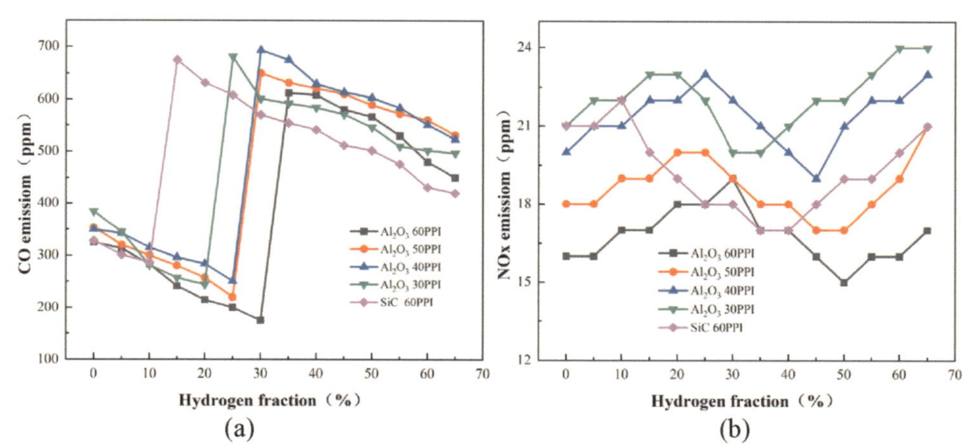

Figure 6.
Variation of pollutant emissions at different HENG. (a) CO (b) NOx.

The initial NOx emission increase stems from enhanced thermal efficiency and elevated combustion temperatures, which accelerate thermal NOx formation from atmospheric N_2 and O_2 [19]. The higher flame temperature of HENG simultaneously reduces prompt NO while increasing thermal NO, with the net effect being dominated by hydrogen promotion of thermal NO formation. The subsequent NOx reduction beyond 35% hydrogen blending occurs as heat transfer shifts from combined flame-radiation to pure flame mode, with flames lifting from the darkened porous media surface, weakening gas preheating and lowering flame temperatures. The final NOx increase at higher hydrogen ratios results from intensified chemical reactions and flame propagation speeds dominating the system. Comparative analysis in **Figure 6** indicates that Al_2O_3@60PPI and SiC@60PPI exhibit the lowest CO emissions, with Al_2O_3@60PPI demonstrating superior NOx reduction. Experimental results demonstrate an inverse relationship between pore density and pollutant emissions for both material types. At equivalent pore densities, Al_2O_3 and SiC show comparable CO emissions, but SiC produces significantly higher NOx levels.

4. Mechanisms of flame instability in porous media combustion of HENG

4.1 Burner model development

A numerical investigation of partially premixed diffusion combustion for HENG within porous media was conducted using Ansys® computational modeling [25]. The simulation geometry was constructed based on precise dimensional measurements of the physical porous media burner depicted in **Figure 7(a)**, while the corresponding numerical representation is illustrated in **Figure 7(b)**. As demonstrated in **Figure 7 (c)**, the burner assembly consists principally of three structural elements: a venturi mixing section, main burner housing, and combustion chamber. Initial fuel-air premixing was accomplished within the venturi section, after which the prepared mixture was conveyed through the burner housing toward the combustion chamber. Subsequent to traversing the porous media section, the gaseous mixture progressed into the reaction zone, visually distinguished by blue coloration in **Figure 7(b)**. System stability was maintained through the introduction of secondary air streams via strategically positioned inlet ports shown in **Figure 7(d)**, thereby optimizing mixture homogeneity throughout the combustion process.

Numerical simulations were conducted for a HENG porous media burner in this study. Existing research findings [26] suggest that the simulation domain should generally extend to approximately six to eight burner head diameters. For the domestic porous media burner structure shown in **Figure 7**, the computational domain was determined to be 6.5 times the burner head diameter, with the geometry illustrated in **Figure 8** as a cylinder measuring 1120 mm in outer diameter and 700 mm in height. The detailed structural dimensions of all burner components are provided in **Table 4**.

4.2 Numerical simulation and boundary definitions

4.2.1 Selection of numerical models

All flow processes are governed by the mass conservation equation, which is expressed as Eq. (2):

$$\nabla \cdot (\rho u) = 0 \tag{2}$$

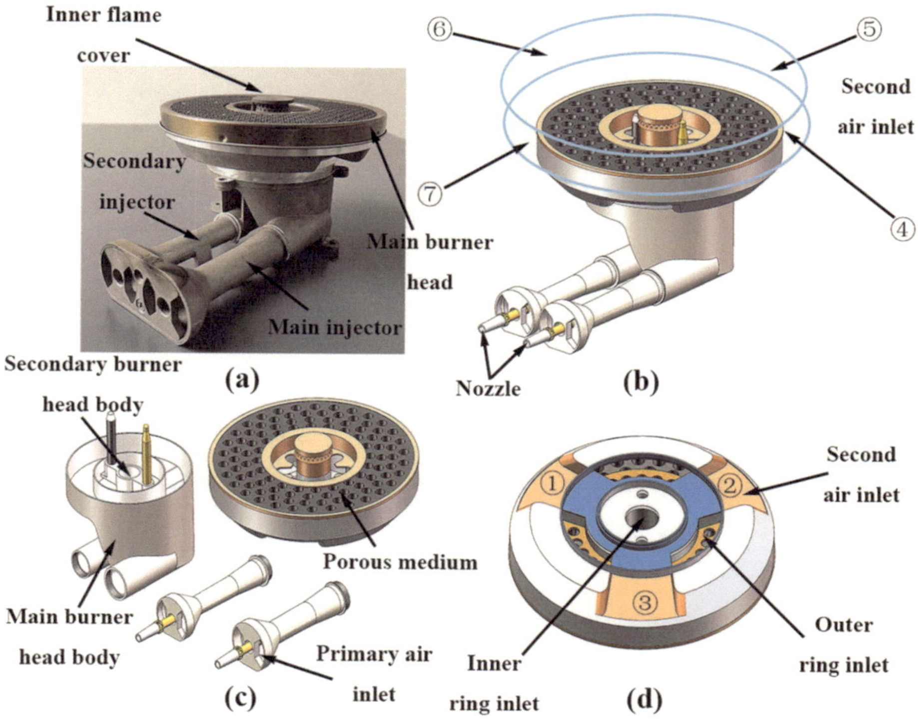

Figure 7.
Porous media gas burner configuration: (a) Physical prototype, (b) Numerical model, (c) Structural components, (d) Main burner head assembly.

ρ—Fluid density, u—Fluid velocity vector.

The energy equation adheres to the law of energy conservation, as expressed in Eq. (3):

$$\frac{\partial(\varepsilon\rho U h)}{\partial x}+\frac{\partial(\varepsilon\rho V h)}{\partial y}=\frac{\partial}{\partial x}\left(k_{eff}\frac{\partial T}{\partial x}\right)+\frac{\partial}{\partial y}\left(k_{eff}\frac{\partial T}{\partial y}\right)+\varepsilon\sum_{i=1}^{N}h_i w_i W_i-\varepsilon\frac{\partial}{\partial X}\left[\sum_{i=1}^{N}Y_i h_i D_{im}\frac{\partial(pgY_i)}{\partial X}\right]-\varepsilon\frac{\partial}{\partial y}\left[\sum_{i=1}^{N}Y_i h_i D_{im}\frac{\partial(pgY_i)}{\partial y}\right] \quad (3)$$

k_{eff}—Effective thermal conductivity, T—Mixture temperature, c—Mole enthalpy of the i-th component, w_i—Mole production rate of the i-th component, Y_i—Mass fraction of the i-th component, D_{im}—Mass diffusion coefficient of the gas.

Thermal radiation transfer through the combustion chamber cannot be neglected in the burner. The discrete ordinates method was adopted for radiation modeling owing to its demonstrated capability to handle various computational conditions, with the corresponding governing relation formulated as Eq. (4):

$$\frac{dI(\vec{r},\vec{s})}{ds}+(a+\sigma_s)I(\vec{r},\vec{s})=an^2\frac{\sigma T_g{}^4}{\pi}+\frac{\sigma_s}{4\pi}\int_0^{4\pi}I(\vec{r},\vec{s})\phi(\vec{s},\vec{s}')d\Omega' \quad (4)$$

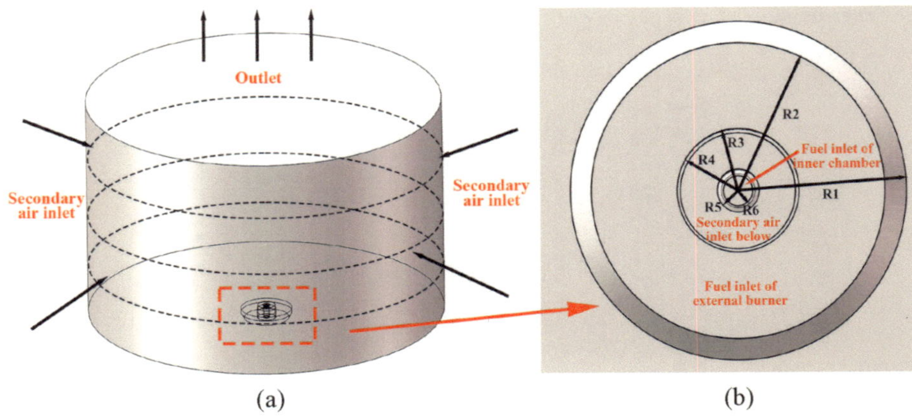

Figure 8.
Burner geometry. (a) Computational domain (b) Structural dimensions.

Nomenclature	Symbols	Radius (mm)	Hydraulic diameter (mm)
Burner radius	R1	86	/
Outer radius of fuel inlet in burner external pipeline	R2	77	96
Inner radius of fuel inlet in burner external pipeline	R3	29	
Outer radius of secondary air inlet below	R4	27.5	35.2
Inner radius of secondary air inlet below	R5	9.9	
Fuel inlet radius in burner internal pipeline	R6	7	14
Secondary air inlet	/	/	
Outer	/	560	1120

Table 4.
Structural categories and components of the burner.

$\vec{r}$—Position vector, $\vec{s}$—Direction vector, $\vec{s}'$—Scattering direction vector, s—Path length, α—Absorption coefficient, n—Refractive index, σ_s—Scattering coefficient, σ—Stefan-Boltzmann constant (5.76×10^{-8} W/(m^{2}K^4)), T_g—Local temperature, I—Radiation intensity, ϕ—Phase function, Ω'—Solid angle.

4.2.2 Boundary conditions

The stabilized combustion characteristics of the gas burner were numerically investigated using steady-state computational analysis. Turbulence effects were characterized through implementation of the Standard k-epsilon approach, complemented by the Eddy-Dissipation Concept for combustion modeling and the Discrete Ordinates method for radiative heat transfer computations. Pressure-velocity coupling was resolved using the SIMPLE scheme, with second-order accurate upwind formulations

Boundary name	Type	Numerical size	Component setting/Mole	Temperature/K
Fuel inlet in burner internal pipeline	Velocity inlet	0.4675 (m/s)	CH_4: 0.1724	300
			O_2: 0.1792	
Fuel inlet in burner external pipeline	Velocity inlet	0.06261 (m/s)	CH_4: 0.1724	300
			O_2: 0.1792	
Secondary air inlet below	Pressure inlet	0 Pa	O_2: 0.2181	300
Secondary air inlet	Pressure inlet	0 Pa	O_2: 0.2181	300
Outlet	Pressure outlet	0 Pa	/	300

Table 5.
Boundary conditions.

employed for discretizing the momentum and energy conservation eqs. A thermal power output of 3.8 kW was specified for the burner system, with the computational domain initialized at 1700 K to ensure proper combustion initiation. The complete set of boundary specifications is documented in **Table 5**.

4.3 Grid independence test and computational model verification

4.3.1 Grid independence test

The three-dimensional model was discretized using hybrid structured-unstructured grids. Local grid refinement was implemented in the porous media zone and burner region, representing the core combustion areas as illustrated in **Figure 9 (a)**. The upper surface of the computational domain, serving as the flue gas outlet, and the peripheral secondary air inlets, located distant from the combustion core, were coarsely meshed while maintaining grid skewness below 0.8. Pure methane was utilized as the fuel with flame temperature along the Y-axis ranging from 0 to 31 mm above the inner flame cap serving as the evaluation criterion. **Figure 9(b)**

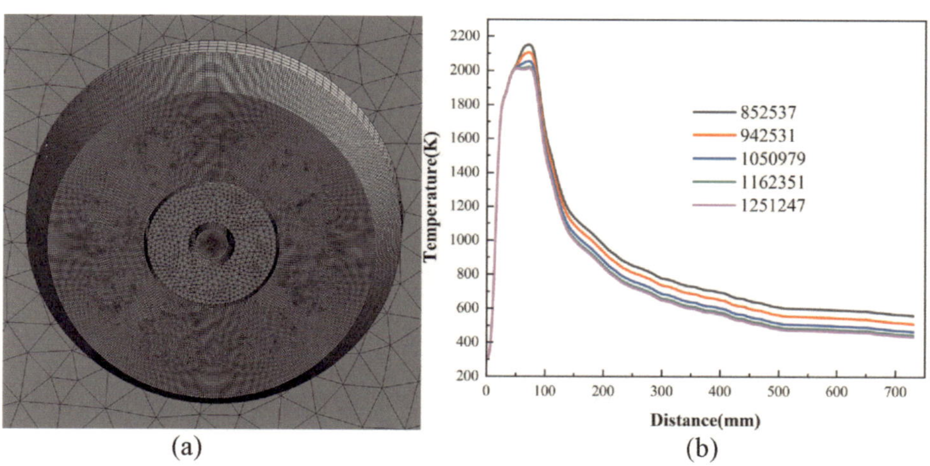

Figure 9.
Mesh generation and independence verification. (a) Localized mesh refinement (b) Grid independence.

demonstrates that at 1.162 million grid elements, the overall grid skewness of 0.78 was achieved with temperature variation constrained within 1%, thereby satisfying computational accuracy requirements. Based on these results, the 1.162-million-element mesh was selected for subsequent numerical analysis.

4.3.2 Model verification

The numerical model predictive capability and parameter configuration appropriateness were evaluated through comparative analysis between simulated and experimental data. Temperature profiles and oxygen concentration distributions were measured along two vertically oriented monitoring lines positioned at Z-coordinates of 51 mm and 71 mm, spanning horizontally from −86 mm to 86 mm, as shown in **Figure 10**. Comparative assessments were performed for both pure methane combustion and methane-hydrogen mixtures containing 30% hydrogen by volume. The comparative results are visually represented in **Figures 11** and **12**.

The numerical simulations exhibit strong correlation with experimental observations for pure methane combustion, as evidenced by the comparative data in **Figure 11**. Temperature profiles and oxygen concentration distributions demonstrate a maximum deviation of 4.1% and mean discrepancy of 1.5% between computational and experimental results. This close correspondence confirms the model ability to reliably predict combustion behavior in porous media burners operating with pure methane. Minor variations between numerical and experimental outcomes primarily stem from differing boundary conditions. The experimental configuration fully open environment facilitated substantial convective heat transfer to surrounding air, while the computational domain relatively confined geometry minimized thermal

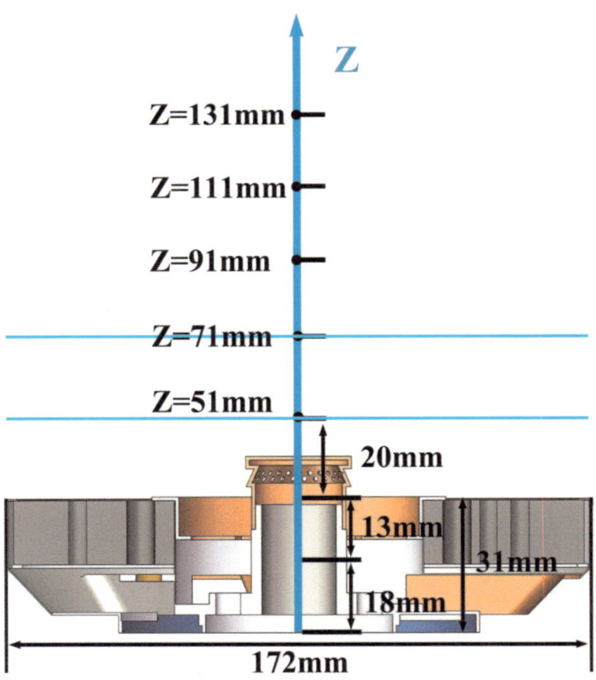

Figure 10.
Schematic diagram of temperature measurement.

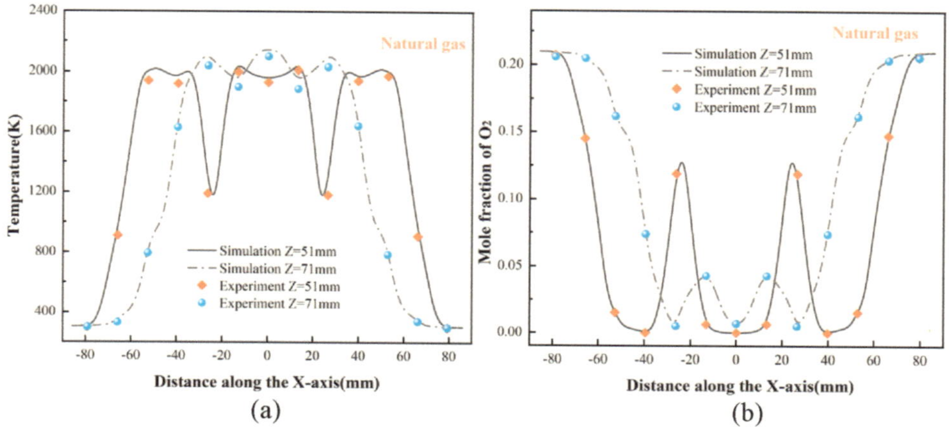

Figure 11.
Validation of numerical model against experimental measurements for pure methane combustion. (a) Temperature (b) Mole fraction of O_2.

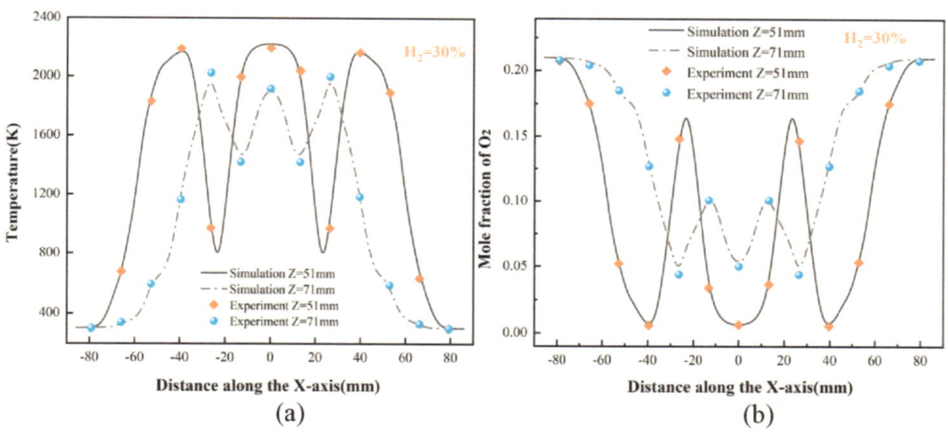

Figure 12.
Validation of numerical model against experimental measurements for 30% hydrogen enriched combustion. (a) Temperature (b) Mole fraction of O_2.

dissipation. These contrasting thermal boundary conditions account for the observed temperature differences at the five measurement locations. The demonstrated predictive accuracy, combined with the physically justifiable nature of observed discrepancies, establishes the numerical model validity for characterizing flame stabilization and combustion performance in the investigated porous media burner system. The model's capability to capture essential combustion features while maintaining computational efficiency renders it suitable for further parametric investigations of methane-hydrogen blended fuel combustion.

4.4 OH radical sensitivity

Methane-hydrogen blended combustion was investigated using a zero-dimensional homogeneous reactor model in Chemkin, based on natural gas-hydrogen combustion

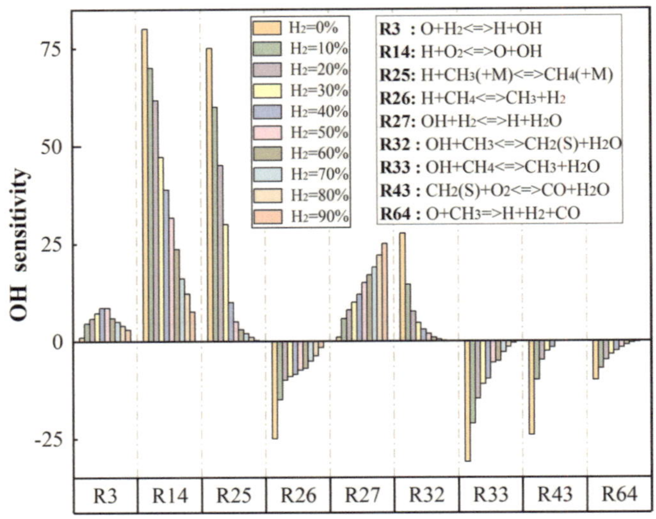

Figure 13.
Reactions sensitivity analysis for OH radicals at different HENG.

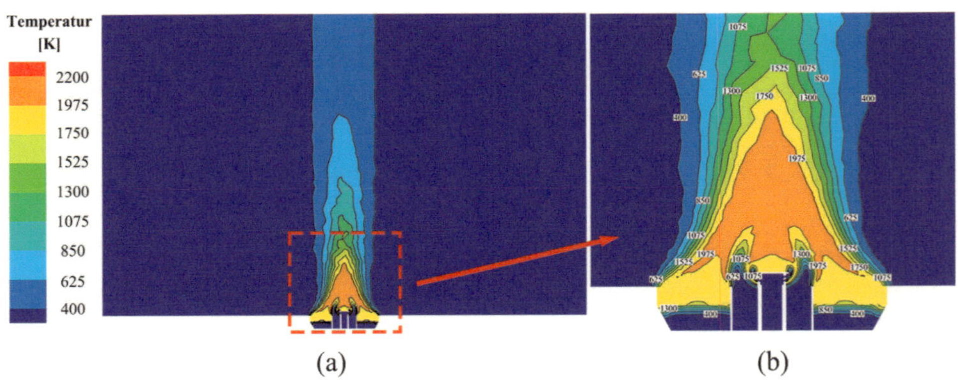

Figure 14.
Temperature distribution under pure methane. (a) Overall schematic diagram (b) Magnified regional visualization.

mechanisms. The GRI-Mech 3.0 mechanism was simplified by identifying key species (CH_4, H_2, O_2, H, O, OH, NO, NO_2, CO, CO_2, and H_2O) and incorporating laminar flame speed data, resulting in a reduced mechanism comprising 19 species and 67 elementary reactions. The validity of this simplified mechanism has been confirmed by Chen et al. [27]. The OH radical represents a crucial reactive species in combustion chemistry, with its generation, depletion, and spatial concentration profile exerting significant influence on HENG combustion dynamics. Increased hydrogen blending ratios have been shown to significantly enhance flame propagation speeds, which are strongly correlated with elevated OH radical concentrations. Moreover, the spatial distribution of OH radicals substantially determines flame stability. Hydrogen addition leads to more concentrated OH radical distributions and narrower flame fronts, thereby improving flame stability. However, excessive concentrations of OH radicals may cause combustion instability and even trigger detonation. Consequently, understanding OH radical distribution is crucial for investigating flame instability in high-

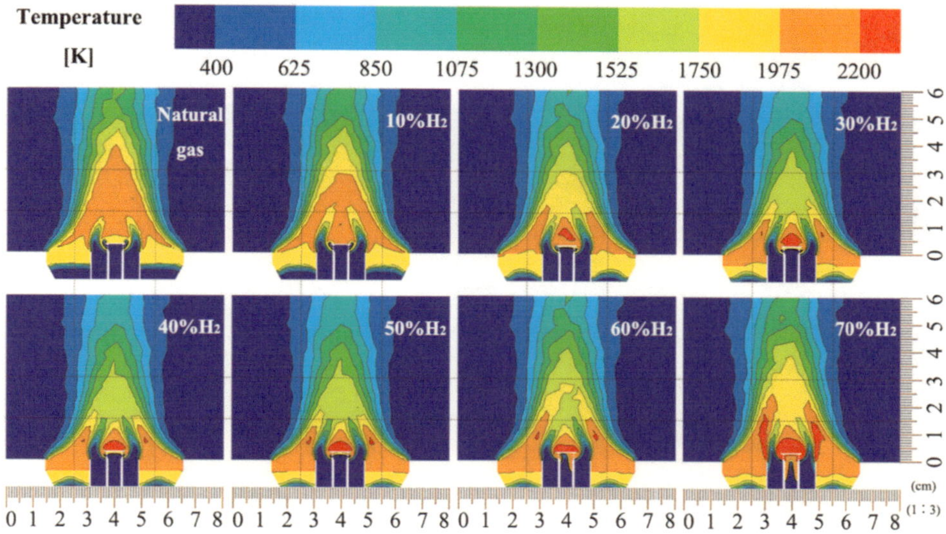

Figure 15.
Temperature distribution contours of Al_2O_3@60PPI porous media burner at different HENG.

hydrogen-blend combustion systems. **Figure 13** presents the variation in OH sensitivity coefficients with different hydrogen blending ratios. Positive sensitivity coefficients indicate that a reaction promotes OH radical formation, while negative values denote inhibitory effects on OH radical production.

Figure 13 demonstrates varying sensitivity patterns in OH radical production as hydrogen content rises in the fuel mixture. The positive influence of elementary reactions R14, R25, and R32 progressively weakens, whereas reaction R27 exhibits significantly strengthened promoting characteristics. Reaction R3 exhibits an initial increase followed by subsequent decrease in its promoting effect. The suppression of OH radical generation by reactions R26, R33, R43, and R64 is progressively attenuated as the hydrogen concentration in the fuel mixture increases. The combustion products H and OH from reaction R14 significantly accelerate reaction rates. As demonstrated by Dagaut et al. [28], R14 plays a dominant role in methane-hydrogen combustion, consuming the largest quantity of H radicals while generating the most OH radicals. Conversely, R27 predominantly consumes OH radicals while producing H radicals. Increased hydrogen concentration promotes the forward reaction, leading to substantial OH consumption and concomitant H radical generation. Below 40% hydrogen blending, reactions R14 and R25 dominate, both consuming H radicals. Beyond this threshold, R25's influence diminishes while R27 becomes increasingly significant. The observed reaction kinetics evolution leads to R14 and R27 becoming the dominant elementary processes, where R14 participates in H radical depletion while R27 contributes to their formation. This alteration in the comparative significance of R14, R25 and R27 at threshold hydrogen blending ratios could potentially trigger combustion regime changes in porous media burners.

4.5 Mechanistic analysis of flame instability in porous media burner

He et al. [29] established that flame structure characteristics can be effectively represented by temperature distribution contours in numerical simulations. **Figure 14**

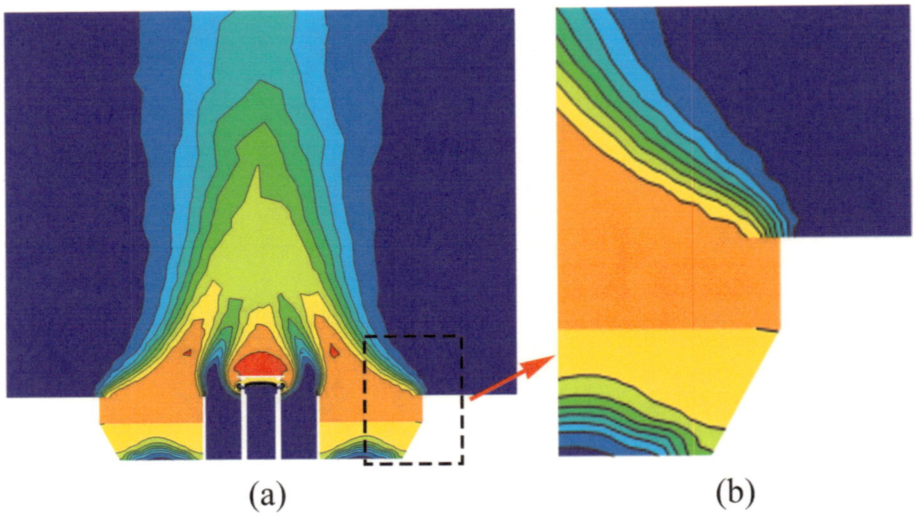

Figure 16.
Temperature contours and local magnification.

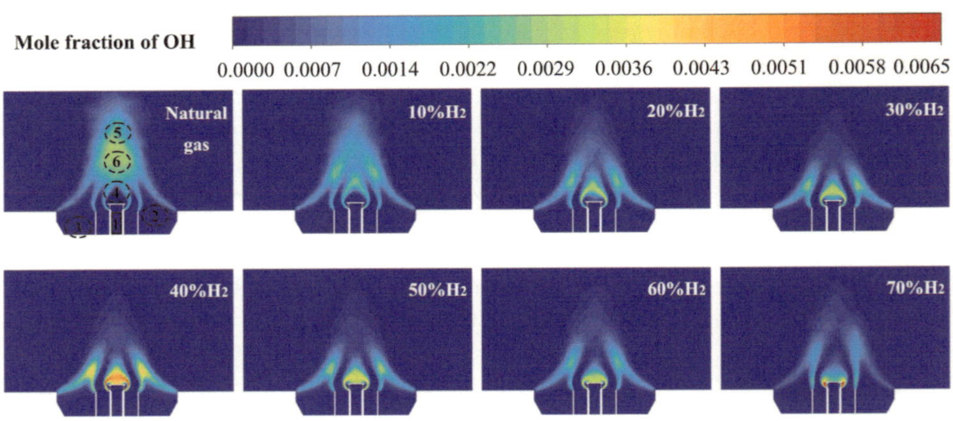

Figure 17.
OH radical concentration distribution contours at different HENG.

illustrates the thermal profile established during steady-state operation of a porous burner fueled exclusively by methane, with **Figure 14(a)** depicting the complete temperature field and **Figure 14(b)** providing an enlarged visualization of the reaction zone. The central flame region is clearly observed to reach the maximum temperature of 1975 K in **Figure 14(b)**, with temperatures progressively decreasing toward the periphery to 850 K in the outermost region. This thermal pattern results from distinct combustion mechanisms. The central flame zone achieves complete methane-oxygen diffusion combustion with continuous atmospheric oxygen mixing, enabling substantial heat release. In contrast, peripheral flame regions exhibit combustion interacting with ambient air, in which convective heat transfer to the flowing air leads to gradual thermal dissipation and thus a lower temperature.

The thermal profiles along the central axial plane (Y = 0) of the Al_2O_3@60PPI porous burner are displayed in **Figure 15** for different HENG compositions. A trend of

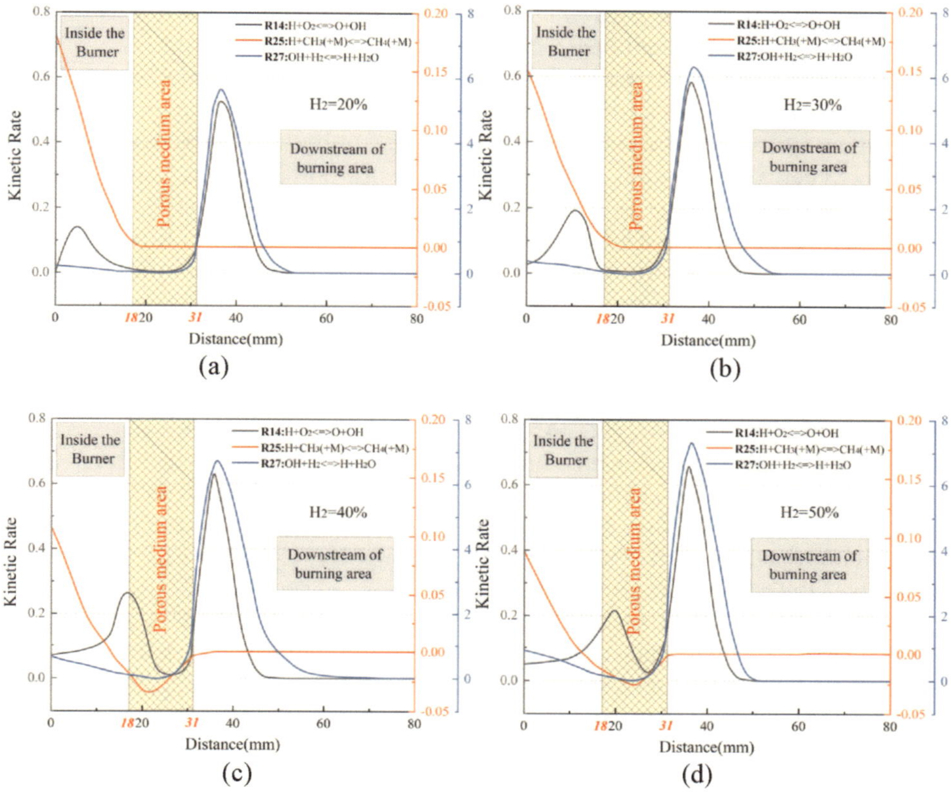

Figure 18.
Hydrogen content effects on critical reaction kinetics of R14, R25, and R27. (a) H_2 = 20% (b) H_2 = 30% (c) H_2 = 40% (d) H_2 = 50%.

initial shortening followed by elongation of flame length is observed with increasing hydrogen addition, in agreement with experimental combustion observations.

To elucidate the fundamental mechanisms responsible for flame behavior modifications in the porous combustion system, a detailed examination of thermal distribution patterns was conducted at 40% hydrogen enrichment. **Figure 16** presents an enhanced visualization of the temperature field, revealing that the maximum flame temperature region has propagated past the porous matrix lower right boundary and infiltrated the internal burner chamber, providing clear evidence of flashback phenomena.

The causes of flashback in high HENG burners were investigated through analysis of elementary OH concentration distribution contours across different HENG compositions. **Figure 17** clearly demonstrates that under pure methane conditions, relatively large flame heights were observed with high OH concentrations predominantly localized in Zones 5 and 6, while other zones maintained lower concentrations. Despite non-uniform OH distribution, no flashback phenomena were detected. With progressive hydrogen enrichment, gradual flame height reduction was accompanied by decreasing peak OH concentrations in Zone 6 and marginally increasing concentrations in Zone 4. A distinct transition occurred at 40% hydrogen blending, characterized by sharply elevated OH concentrations in Zones 4 and 6 reaching maximum values, accompanied by substantially amplified interzonal concentration differentials. This phenomenon indicates that the OH radical concentration distribution has become significantly heterogeneous between the inner and outer flame zones at this stage. Further hydrogen enrichment beyond 40% resulted in declining peak OH concentrations in Zones 4 and 6 correlating with increasing flame height. At 70% hydrogen blending, penetration of high-temperature flame from Zone 4 into the burner inner cavity was observed, where intensive combustion reactions depleted available oxygen and triggered dramatic OH concentration increases near the inner chamber's jet orifices [30].

The examination of **Figure 13** indicates that elementary reactions R14 ($H + O_2 \rightleftharpoons O + H_2$), R27($OH + H_2 \rightleftharpoons H + H_2O$), and R25($H + CH_3(+M) \rightleftharpoons CH_4(+M)$) significantly influence high-energy gas combustion dynamics. To elucidate the abrupt variation in OH radical concentration, the kinetic behavior of these key reactions was systematically evaluated across varying hydrogen enrichment levels. As evidenced by **Figure 18**, the kinetic rates of R14 and R27 display a consistent enhancement with increasing hydrogen content, whereas R25 exhibits a gradual reduction in reaction activity. Under elevated hydrogen concentrations, R14 demonstrates accelerated kinetics within the burner chamber, particularly near the porous matrix lower boundary. Concurrently, R25 within the porous medium undergoes a transition from negligible activity to negative rate values, signifying inhibition of the forward pathway and activation of the reverse process. Parallel observations reveal that R27 reaction intensity in the porous region increases from an initially undetectable level. During stable combustion at 30% hydrogen content, measurable R25 activity is confined to the porous media lower surface vicinity, while R14 and R27 remain kinetically inactive in this zone. At 40% hydrogen enrichment, peak reaction rates within the porous medium are quantified as: R14 (0.27), R27 (0.30), and R25 (−0.03). This condition corresponds to R25 forward reaction suppression coupled with reverse reaction promotion, yielding significant hydrogen radical production. The consequent elevation in H radical concentration preferentially drives R14 forward kinetics, resulting in substantial OH radical generation, in consistent with the findings of Wang et al. [31].

The dramatic OH concentration increase at 40% hydrogen blending is consequently explained by synergistic interactions among R14, R25, and R27 within the porous medium: while R14 and R27 predominantly proceed in the forward direction, R25 preferentially undergoes reverse transformation. The collective enhancement of R14/R27 forward rates and R25 reverse rate directly accounts for the observed OH sudden surge.

5. Conclusions

A novel porous media gas burner was developed in this work for stable operation with HENG at elevated blending ratios. Through combined experimental investigation and numerical simulation, the influence of porous media materials and structural parameters on combustion characteristics was investigated under different HENG. The transformation mechanism of combustion modes under high hydrogen proportions was theoretically analyzed. Furthermore, Fundamental mechanisms underlying flashback occurrences in gas burner were elucidated, establishing essential theoretical frameworks and technical foundations for deploying high-ratio HENG in domestic gas appliances. The following conclusions were drawn:

1. With increasing hydrogen blending ratios, a transition in combustion modes is observed in porous media burners, which directly affects the thermal efficiency

and CO/NOx emissions of HENG. Both thermal efficiency and NOx emissions increase with higher hydrogen content before and after the combustion mode transition, while CO emissions progressively decrease. When the critical hydrogen blending ratio is reached, the combustion mode transition triggers a rapid increase in CO emissions accompanied by decreasing trends in both thermal efficiency and NOx emissions. The critical hydrogen ratio for this transition is strongly dependent on the material and structural parameters of the porous media. Comparative analysis reveals that SiC exhibits inferior overall performance to Al_2O_3, with higher pore densities enabling larger critical hydrogen ratios. The maximum allowable hydrogen blending ratio for Al_2O_3@60PPI is determined to be 35%.

2. Progressive infiltration of reactions R25($H + CH_3(+M) \rightleftharpoons CH_4(+M)$), R27 ($OH + H_2 \rightleftharpoons H + H_2O$) and R14($H + O_2 \rightleftharpoons O + H_2$) into the porous structure was observed at 40% hydrogen blending. The forward reaction rates of R14 and R27 were enhanced, while R25 predominantly promoted the reverse reaction. This synergistic effect characterized by increased forward rates (R14/R27) and accelerated reverse rate (R25) induced an abrupt surge in OH radical generation, ultimately leading to flame instability and flashback phenomena.

Acknowledgements

This work is supported by the National Key R&D program of China (No. 2021YFB4001604) and the Natural Science Foundation of Hunan Province (No. 2025JJ60326) and the Industrial Guiding Project of Fujian Province of China (No. 2023H0054).

CRediT authorship contribution statement

Zhongmin Wan: Conceptualization, Methodology, Writing – review & editing. **Jie Niu:** Conceptualization, Writing – review & editing, Investigation. **Yiyu Chen:** Conceptualization, Methodology, Supervision, Funding acquisition, Project administration. **Chen Yang:** Supervision, Project administration.

Author details

Chen Yang[1], Jie Niu[1], Yiyu Chen[2] and Zhongmin Wan[2]*

1 Fujian Universities Engineering Research Center of Reactive Distillation Technology, College of Chemical Engineering, Fuzhou University, Fuzhou, Fujian, China

2 College of Energy and Electrical Engineering, Hunan Institute of Science and Technology, Yueyang, China

*Address all correspondence to: zhongminwan@hotmail.com

References

[1] Abdin Z, Zafaranloo A, Rafiee A, et al. Hydrogen as an energy vector. Renewable and Sustainable Energy Reviews. 2020;**120**:109620

[2] Zhu J, Wang S, Pan J, et al. Experimental study on leakage temperature field of hydrogen blending into natural gas buried pipeline. Applied Energy. 2024;**359**:122656

[3] Cheng L, Li W, Peng S, et al. Study of combustion characteristics of hydrogen-doped natural gas in industrial boilers. International Journal of Hydrogen Energy. 2024;**92**:590-604

[4] Meziane S, Bentebbiche A. Numerical study of blended fuel natural gas-hydrogen combustion in rich/quench/lean combustor of a micro gas turbine. International Journal of Hydrogen Energy. 2019;**44**(29):15610-15621

[5] Sun M, Huang X, Hu Y, et al. Effects on the performance of domestic gas appliances operated on natural gas mixed with hydrogen. Energy. 2022;**244**:122557

[6] Fang Z, Zhang S, Huang X, et al. Performance of three typical domestic gas stoves operated with methane-hydrogen mixture. Case Studies in Thermal Engineering. 2023;**41**:102631

[7] Zhao Y, Mcdonell V, Samuelsen S. Influence of hydrogen addition to pipeline natural gas on the combustion performance of a cooktop burner. International Journal of Hydrogen Energy. 2019;**44**(23):12239-12253

[8] Chen Y, Long L, Niu J, et al. Investigation on the combustion characteristics and environmental effects of hydrogen-doped natural gas for domestic water heaters. ACS Omega. 2023;**8**(50):48370-48380

[9] Choudhury S, McDonell V, Samuelsen S. Combustion performance of low-NOx and conventional storage water heaters operated on hydrogen enriched natural gas. International Journal of Hydrogen Energy. 2020;**45**(3):2405-2417

[10] Sorgulu F, Ozturk M, Javani N, et al. Experimental investigation for combustion performance of hydrogen and natural gas fuel blends. International Journal of Hydrogen Energy. 2023;**48**(88):34476-34485

[11] Pan J, Wu D, Liu Y, et al. Hydrogen/oxygen premixed combustion characteristics in micro porous media combustor. Applied Energy. 2015;**160**:802-807

[12] Pantangi V, Mishra S, Muthukumar P, et al. Studies on porous radiant burners for LPG cooking applications. Energy. 2011;**36**(10):6074-6080

[13] Trimis D, Durst F. Combustion in a porous media compact heat exchanger unit-experiment and analysis. Experiment Thermal and Fluid Science. 2004;**28**:183-192

[14] Mital R, Gore JP, Viskanta R. A study of the structure of submerged reaction zone in porous ceramic radiant burners. Combustion & Flame. 1997;**111**(3):175-184

[15] Bouma P, De Goey L. Premixed combustion on ceramic foam burners. Combustion and Flame. 1999;**119**(1–2):133-143

[16] Laphirattanakul P, Laphirattanakul A, Charoensuk J. Effect of self-entrainment and porous geometry on stability of premixed LPG

porous burner. Applied Thermal Engineering. 2016;**103**:583-591

[17] Herrera B, Cacua K, Olmos-Villalba L. Combustion stability and thermal efficiency in a porous media burner for LPG cooking in the food industry using Al_2O_3 particles coming from grinding wastes. Applied Thermal Engineering. 2015;**91**:1127-1133

[18] Hariharan V, Mishra DP. Dynamic flame stability diagnosis of inverse jet flame using CH^* chemiluminescence. Fuel. 2021;**285**:119277

[19] Liu X, Zhu G, Asim T, et al. Combustion characterization of hybrid methane-hydrogen gas in domestic swirl stoves. Fuel. 2023;**333**:126413

[20] Chen Y, Niu J, Liu W, et al. Experimental analysis and modeling on the blending limit of domestic burner with porous media for hydrogen enriched natural gas. International Journal of Hydrogen Energy. 2024;**88**:1321-1331

[21] Muthukumar P, Shyamkumar PI. Development of novel porous radiant burners for LPG cooking applications. Fuel. 2013;**112**:562-566

[22] GB/T 13611-2018. Classification and basic characteristics of city gas

[23] Wang Y, Han H, Sohn C. A comparative study of chemical-kinetic mechanisms for combustion of methane/hydrogen/air mixtures. International Journal of Aeronautical and Space Sciences. 2023;**25**(2):1-21

[24] Sun X, Liu H, Duan X, et al. Effect of hydrogen enrichment on the flame propagation, emissions formation and energy balance of the natural gas spark ignition engine. Fuel. 2022;**307**:121843

[25] ANSYS, Inc. ANSYS Fluent User's Guide, Release 22R1; 2022

[26] Deymi-Dashtebayaz M, Rezapour M, Sheikhani H, et al. Numerical and experimental analyses of a novel natural gas cooking burner with the aim of improving energy efficiency and reducing environmental pollution. Energy. 2023;**263**:126020

[27] Chen Y, Long L, Hu J, et al. The combustion characteristics and stable limit of a novel combustor with gradient porous media for hydrogen-enriched natural gas. Journal of the Energy Institute. 2024;**116**:101743

[28] Dagaut P, Dayma G. Hydrogen-enriched natural gas blend oxidation under high-pressure conditions: Experimental and detailed chemical kinetic modeling. International Journal of Hydrogen Energy. 2006;**31**(4):505-515

[29] He J, Chen Z, Jiang X, et al. Combustion characteristics of blast furnace gas in porous media burner. Applied Thermal Engineering. 2019;**160**: 113970

[30] Zhan X, Chen Z, Qin C. Effect of hydrogen-blended natural gas on combustion stability and emission of water heater burner. Case Studies in Thermal Engineering. 2022;**37**:102246

[31] Wang F, Xiao C, Zhang D, et al. The effect of CH_4 addition on high-H_2 syngas combustion characteristics under N_2/CO_2 dilution. Journal of the Energy Institute. 2024;**114**:101600

Chapter 8

Supercritical CO_2-Processed Polymeric Foams for Process Intensification Structure, Functionality, and Advanced Applications

Ignacio García-Casas, Antonio Montes, Ludisbel León-Marcos, Clara Pereyra and Diego Valor

Abstract

Polymeric materials foamed using supercritical carbon dioxide ($scCO_2$) offer a sustainable and versatile basis for developing advanced porous structures with highly adjustable morphology, density, and functionality. This chapter provides a comprehensive overview of the physical principles and processing strategies involved in generating foams with hierarchical porosities, emphasizing their potential for process intensification. The correlations between structure and functionality are presented and illustrated using case studies in which porosity values of over 90%, open-cell contents of up to 96%, and functional loads of over 18% are achieved in $scCO_2$-assisted impregnation processes. The advantages of this technology for the *in-situ* formation of catalysts are analyzed, demonstrating increases in hydrogen production of up to ~60% compared to traditional packed beds. Integrating functional impregnation and foaming in a single step, together with using a clean, recyclable physical agent such as $scCO_2$, makes these foams ideal for advanced applications in compact reactors, controlled release systems and biomedical platforms. The combination of sustainability, multiscale morphological control and multifunctional properties establishes these foams as essential components in the development of more efficient, intensified, and environmentally responsible emerging technologies.

Keywords: polymeric foams, porosity, intensification, $scCO_2$ foaming, impregnation

1. Introduction

Modern chemical engineering is facing an increasing number of challenges that demand a fundamental re-evaluation of the way processes are designed and operated. The need to optimize efficiency, reduce the environmental footprint, and accelerate product development has led to the development of process intensification (PI). Within this framework, advanced materials—particularly polymeric foams—and innovative processing technologies, such as the use of $scCO_2$, have emerged as key drivers of industrial transformation. This chapter will explore how the combination of the unique structural properties of polymeric foams and the environmentally friendly processing capabilities of supercritical carbon dioxide contribute to PI objectives, with a particular focus on functionalizing these materials for advanced applications.

1.1 Process intensification in modern chemical engineering: Concepts and objectives

Process intensification has established itself as a revolutionary design approach in chemical engineering. Its primary purpose is to significantly reduce the size of process equipment and plants while substantially improving the overall efficiency and sustainability of operations. This paradigm goes beyond the incremental optimization of existing processes, seeking a fundamental restructuring that maximizes the effectiveness of intra- and intermolecular events. A central tenet of PI is ensuring that each molecule undergoes uniform processing by optimizing driving forces and resistances at all scales and maximizing the specific surface areas where these interactions occur. Unlike traditional engineering design, which often approaches process stages sequentially and in isolation, PI promotes an integrative approach, combining chemical and physical phenomena to achieve operational synergies. The concept was formalized by Stankiewicz and Moulijin in 2000, who described it as a 'toolbox' encompassing novel equipment and innovative methods to achieve these goals [1].

The reasons for adopting PI are multifaceted and respond to market pressures and growing demands for sustainability. In the biopharmaceutical sector, for instance, PI is vital for accelerating drug development, enhancing efficiency and productivity in line with strict Good Manufacturing Practice (GMP) standards, and improving operational flexibility to accommodate diverse therapies. Specific PI objectives include boosting productivity, reducing the cost of goods, expediting development and commercialization, and achieving greater agility in manufacturing operations. From a sustainability perspective, PI enables radical reductions in energy consumption and waste generation, resulting in technologies that are more economical, efficient, and environmentally friendly. Speed to market has become particularly important for a product's competitive position and price, highlighting the strategic importance of PI in responding to market dynamics. Sustainability is not just an ethical objective, but a necessity imposed by increasingly stringent regulations and growing corporate awareness of environmental image [2].

Process intensification is therefore a driver of industrial transformation that goes beyond mere technical optimization. By adopting PI, companies are not only seeking to improve their existing processes; they are also redefining the way they conceive and execute their chemical engineering projects. This shift is moving them toward more agile, adaptable, and responsible production. This paradigm shift is evident in the increasing popularity of continuous processes, also known as flow chemistry, as opposed to traditional batch processes. Continuous systems offer advantages in terms of scalability, safety, and efficiency. Examples of this evolution include perfusion bioreactors and single-use systems in biomanufacturing, which enable a higher throughput of individual therapies and provide unparalleled flexibility in adapting to different production methods within the same facility [3].

Another significant advance is reactive distillation, a technology that combines chemical reactions and separations in a single process, resulting in substantial energy savings and reduced CO_2 emissions. Membrane engineering and the use of immobile mixers are additional examples of PI methods and equipment that improve separation efficiency and mixing, respectively. This demonstrates how PI innovation often occurs at the intersection of design and process engineering [1]. The true power of PI lies in the co-evolution and integration of equipment and methods. Having innovative equipment is not enough if it is not accompanied by an optimized method of operation, and vice versa. This interdependence highlights the importance of a holistic approach to research and development, where equipment design and process optimization are developed together to maximize the benefits of PI.

1.2 The role of porous materials and polymeric foams in process intensification

Polymeric foams are materials composed of a mixture of polymers and gases, giving them a distinctive, porous, cellular structure. These materials are also known as cellular, blown, or expanded polymers. Their characteristic properties include low density, high strength-to-weight ratio, excellent thermal insulation, and remarkable energy absorption capacity. The most common polymers used in foam production are polystyrene, polyurethane, natural rubber, and polyvinyl chloride (PVC). Polymeric foams can be classified according to various criteria, such as hardness (elastic, flexible or rigid), density (high, medium or low foaming), and cell structure (open or closed). The distinction between open-cell and closed-cell foams is functionally critical: open-cell foams allow fluids to pass through their interconnected structure, while closed-cell foams prevent water absorption, making them ideal for buoyancy or insulation applications [4].

The porous structure of polymeric foams makes them intrinsic enablers of process intensification. The vast internal surface area they provide is crucial for enhancing mass and heat transfer in chemical processes, which aligns directly with one of the fundamental principles of PI: maximizing specific surface areas. Engineering these porous structures involves more than just creating a lightweight material; it also involves sculpting their microstructure (cell size, interconnectivity and porosity) to optimize transport phenomena. The ability to control these morphological characteristics is central to PI foam research and development, transforming morphology into a critical design variable for process efficiency.

One area of particular interest is the potential of hybrid polymeric foams for multiscale functionality. Hybrid metal foams, for example, consisting of a thin metal film on a porous polymer substrate, have been shown to improve mechanical and electrical properties while maintaining the flexibility of the polymer matrix [5]. This synergy mitigates the weaknesses of the individual materials (e.g., metals are stiff, but polymers are flexible).

Among their numerous applications, open-cell polymeric foams have shown promise as supports in catalytic systems due to their high specific surface area and excellent mass transfer properties. A detailed discussion of these catalytic applications is provided in Section 4.1.

1.3 $scCO_2$ as a green processing tool for polymers

Supercritical carbon dioxide is a unique state of matter in which the liquid and gas phases become indistinguishable due to the pressure and temperature being above their critical values (Tc = 31.1°C and Pc = 7.38 MPa). In this state, $scCO_2$ exhibits intermediate properties between a gas and a liquid, possessing a density similar to a liquid but with diffusivity and viscosity comparable to a gas. One of its most notable characteristics is its ability to 'tune' its solvent power and transport properties by making precise adjustments to the temperature and pressure. In addition to these adjustable physical properties, supercritical carbon dioxide has intrinsic advantages that make it an ideal solvent: it is inert, nontoxic, non-flammable, inexpensive, widely available, fully recyclable, and leaves no residue in the final product [6].

By replacing petroleum-derived solvents such as hexane, ethanol, and acetone, $scCO_2$ drastically reduces health and environmental risks by eliminating the production of effluents and pollutants. This feature aligns it directly with the principles of green chemistry and sustainable manufacturing practices [6]. In the polymer foaming process, $scCO_2$ is injected directly into the molten polymer, eliminating the need for chemical blowing agents, which are often toxic and harmful to the environment [7]. This improves the safety of the process and contributes to a cleaner end product.

One of the most significant advantages of $scCO_2$ is its ability to provide improved control over foam properties. It enables the precise adjustment of foaming parameters, resulting in foams with a smaller pore size, higher pore density, and superior performance in a variety of applications [7]. In addition, $scCO_2$ acts as an effective plasticizer for polymers, reducing their viscosity and glass transition temperature (Tg). This facilitates polymer processing and improves the diffusion of additives within the matrix [8]. The versatility of supercritical carbon dioxide extends to a wide range of industries, including food, pharmaceuticals, perfumery, automotive, footwear, medical devices, and composite materials.

The ability to adjust the properties of $scCO_2$ in real time by controlling pressure and temperature is a fundamental feature that enables high-precision materials engineering. In the context of foaming, for example, this 'tunability' enables the precise control of properties such as pore size and density [9]. In impregnation, it directly affects the solubility of the active ingredient and its ability to diffuse into the polymer. It can even modify the crystallinity and thermal properties of the polymer [10]. This tunability establishes $scCO_2$ as a precision engineering tool at the molecular and micrometer levels. It is essential for fabricating polymeric foams with highly customized morphologies and functionalities, which is critical for advanced applications requiring very specific properties, such as tissue engineering scaffolds with controlled porosity and connectivity [11] or drug delivery systems with precise release profiles [12].

2. Fundamentals of polymeric foam formation with $scCO_2$

2.1 Phase behavior and solubility

CO_2 acts as a physical blowing agent, and its interaction with the polymer dictates the entire cell formation and structure. At this point, it is necessary to understand and control the foaming process. Moreover, in the case of $scCO_2$, it can rapidly penetrate and diffuse into the polymer matrix, facilitating quick saturation times and providing enough gas for foam expansion. In addition, $scCO_2$ properties as density, viscosity, and thus solvent power, can be modulated by varying temperature and pressure, allowing for fine-tuning of the foaming process and resulting foam morphology.

The solubility of CO_2 in polymers generally increases with increasing pressure at a constant temperature, up to a certain point [13]. Higher pressure means more CO_2

molecules are forced into the polymer, leading to a higher concentration of dissolved gas. However, the relationship between temperature and solubility is more complex. Generally, at constant pressure, an increase in temperature can initially increase solubility due to increased polymer free volume. However, at higher temperatures, the decreased density of $scCO_2$ and reduced interaction strength can lead to a decrease in solubility [14]. The optimal foaming temperature window is often a balance between sufficient solubility for gas loading and appropriate polymer viscosity for cell growth and stability.

Another point to consider is the chemical nature of the polymer which significantly influences CO_2 solubility. Polymers with polar groups often exhibit stronger interactions with CO_2, leading to higher solubility compared to nonpolar polymers [10]. For instance, CO_2 generally has high solubility and diffusivity in polymers, especially those with polar groups, providing more gas for cell growth.

The solubility of CO_2 into the polymer is directly related with the plasticization effect. In this sense dissolved CO_2 acts as a plasticizer for the polymer, reducing its glass transition temperature and viscosity. This plasticization is essential for allowing the polymer to deform and expand during cell growth. The extent of Tg depression is directly related to the amount of dissolved CO_2 [15].

Understanding the phase equilibrium of the polymer-CO_2 system is crucial for predicting and controlling the foaming process. Phase diagrams (e.g., pressure-composition diagrams) can illustrate the conditions under which CO_2 is dissolved in the polymer, the onset of phase separation (nucleation), and the conditions for foam stability. One method to elaborate a phase equilibrium diagram is using a view cell equipment. The polymer-CO_2 mixture is directly observed in a high-pressure cell equipped with sapphire windows. Pressure and/or temperature are adjusted until turbidity (the "cloud point") is observed, indicating phase separation (either the formation of CO_2 bubbles within the polymer or the precipitation of the polymer).

2.2 Nucleation, cell growth, and stabilization

Nucleation is the birth of gas bubbles within the polymer matrix. It is the initial, crucial step that dictates the number of cells (cell density) in the final foam. The prerequisite for nucleation is supersaturation. This occurs when the dissolved CO_2 in the polymer is suddenly pushed out of its equilibrium state. In $scCO_2$ foaming, this is typically achieved by a rapid pressure drop (decompression) or, less commonly, a rapid temperature increase. The sudden reduction in solubility means the polymer now holds more CO_2 than it can at the new conditions [16].

For a new bubble to form, it must overcome an energy barrier. This barrier arises from the work required to create a new gas-polymer interface (surface tension) and the work needed to expand the polymer to accommodate the bubble. According to classical nucleation theory, a bubble becomes stable and grows when it reaches a "critical radius." Bubbles smaller than this critical size tend to collapse, while those larger will grow.

On the other hand, nucleation can be homogeneous or heterogeneous. Homogeneous nucleation occurs spontaneously within the bulk of the polymer-CO_2 solution. It requires a very high degree of supersaturation and is less common in practical foaming processes. However, heterogeneous nucleation is the dominant mechanism in most $scCO_2$ foaming processes. It occurs at pre-existing imperfections (e.g., impurities, internal voids, or intentionally added nucleating agents like nanoparticles) within the polymer. These sites lower the energy barrier for bubble formation, making it easier for nuclei to form. Rough surfaces also provide such sites.

Anyway, nucleation can be influenced by factors such as depressurization rate. A faster depressurization rate leads to a higher degree of supersaturation, resulting in a higher nucleation rate and thus more, smaller cells [17]. The initial concentration of CO_2 (generally the higher initial concentration of CO_2 the higher the nucleation) in the polymer and its plasticization level, both of which also influence the energy barrier for nucleation and this fact can be tuned by saturation pressure and temperature [18].

Moreover, polymer properties, such as viscosity, glass transition temperature, and interfacial tension with CO_2, must be taken into account [19]. For instance, lowers glass transition and viscosity, facilitating nucleation. Once nuclei are formed, they begin to grow by absorbing more CO_2 from the surrounding supersaturated polymer matrix. Cell growth is primarily driven by the diffusion of dissolved CO_2 from the polymer into the existing bubbles. The concentration gradient between the supersaturated polymer and the growing bubble drives this mass transfer.

As CO_2 continues to diffuse out of the polymer into the bubbles, the polymer matrix around the bubbles loses its plasticizing agent, leading to an increase in its viscosity. This viscosity increase eventually slows down and stops cell growth. Higher temperatures generally increase the diffusivity of CO_2 in the polymer and lower the polymer's viscosity, leading to faster and larger cell growth. However, excessively high temperatures can lead to cell coalescence.

In this step, the melt strength and viscosity of the polymer-gas mixture are critical [20]. A polymer with sufficient melt strength can resist the internal pressure of the growing bubbles without rupturing, allowing for uniform growth. In this sense, too low a viscosity can also lead to cell coalescence.

Once bubbles are formed, stabilization is needed, as the process of freezing the cellular structure to prevent cell collapse, shrinkage, or further unwanted growth and coalescence after the desired foam morphology is achieved. The most common and effective stabilization method is rapid cooling of the foamed polymer below its glass transition temperature for amorphous polymers, or below its crystallization temperature for semi-crystalline polymers. This causes the polymer to solidify, locking the cell structure in place. As the polymer cools, its viscosity rapidly increases, hindering further CO_2 diffusion, restricting cell growth, and preventing cell coalescence or collapse. As a faster cooling rate generally leads to better preservation of the fine cell structure formed during nucleation and growth, a polymer with a suitable Tg and ability to solidify quickly is ideal. Thicker cell walls provide more structural integrity and resistance to collapse.

In essence, successful $scCO_2$ foaming requires a precise control over the interplay between the thermodynamics of CO_2 solubility and the kinetics of nucleation, cell growth, and solidification to achieve a desired foam morphology with optimal properties.

2.3 Strategies for controlling cell interconnectivity and open-cell content to generate porosity

Controlling cell interconnectivity and the proportion of open cells is crucial for specific applications such as tissue engineering, filtration, and insulation, where mass transport properties are quite important [21]. Achieving precise control often requires advanced techniques beyond standard foaming parameters. Traditional single-step foaming processes offer limited control over the intricate details of cellular architecture, particularly interconnectivity. Advanced foaming techniques represent a higher level of control, enabling the creation of complex, hierarchical cellular structures.

2.3.1 Multistage depressurization

This technique allows for the creation of a wider range of pore sizes and can yield bimodal or even trimodal cellular structures. The process typically involves a slower initial depressurization step, which facilitates the formation of larger pores. This is then followed by a rapid second depressurization step, which triggers the nucleation of new, smaller pores and promotes the additional growth of existing ones [22]. For example, in poly(ε-caprolactone) (PCL) scaffolds, this method has successfully produced large pores (over 100 μm) from the slow step and smaller pores (below 40 μm) from the fast step [11]. This approach moves beyond optimizing individual parameters to designing sequences of operations, offering refined control over pore size distribution.

2.3.2 Gas counter-pressure molding (GCP)

GCP is an injection molding technique where the mold cavity is prefilled with a high-pressure inert gas before the polymer-SCF melt is injected. This pressure is maintained throughout the filling stage. The counter-pressure delays bubble nucleation during the filling stage, which not only improves the surface quality of the molded part but also helps achieve more uniform foaming throughout the material [23]. Furthermore, GCP can be leveraged to increase foam density and promote bubble coalescence between layers, which is beneficial for enhancing gas permeability in applications like filtration.

2.3.3 Temperature-induced foaming

This foaming approach begins by saturating the polymer with a supercritical fluid at lower temperatures and pressures. Foaming is then induced by quickly transferring the saturated polymer to a hot medium (e.g., hot water or oil). The sudden temperature increase significantly reduces the blowing agent's solubility (above the polymer's depressed glass transition), prompting rapid cell nucleation and growth [24].

These multistage processes represent a significant advancement in achieving hierarchical control over foam structures. They enable the creation of complex architectures that would be difficult or impossible to achieve with single-step methods. This implies that advanced manufacturing paradigms are necessary for producing highly customized foam architectures with tailored interconnectivity.

2.3.4 Post-foaming treatments

While in-process parameters establish the initial foam structure, post-foaming treatments offer a "second chance" to refine or drastically alter the cellular morphology, particularly the interconnectivity. These methods are crucial for applications demanding high levels of pore interconnectivity.

2.3.5 Ultrasonication

This technique involves applying external ultrasonic vibrations to the foamed polymer, significantly improving cell interconnection and permeability. It offers a solvent-free method for creating open-cell structures, overcoming the typically closed-cell morphology obtained with CO_2 batch foaming [25]. The mechanism involves the application of instantaneous energy through acoustic vibration, micro-jets, and ultrasonic cavitation, which generates localized high temperatures and pressures that break cell walls, thereby increasing permeability. If applied at the beginning of cell nucleation, ultrasonication can enhance nucleation and lead to higher cell density. If applied after cell nucleation and growth, it results in cell wall rupture, yielding a more open and interconnected cellular structure.

2.3.6 Polymer leaching

This method involves mixing a water-soluble sacrificial polymer (like poly(ethylene oxide) (PEO) or poly(vinyl alcohol) (PVA)) or particulate porogens (such as sodium bicarbonate ($NaHCO_3$), silica, or sodium chloride (NaCl)) with the main polymer. After foaming, the sacrificial material is washed away with water, leaving behind a highly porous and interconnected structure. For example, adding microparticulate silica has been shown to increase the interconnectivity in polylactide scaffolds [26]. In another instance, PCL/cellulose nanofiber composites, where $NaHCO_3$ acted as a porogen, achieved an impressive 82% open-cell content.

2.3.7 Annealing

In batch foaming, an isothermal thermal treatment, or annealing, is often used. This step ensures the blowing agent fully saturates the polymer and helps perfect its crystal structure. During annealing, small, imperfect crystals can mature into more stable, tightly packed, and flawless crystalline forms. This ultimately affects the foam's overall shape and stability.

2.3.8 Stretching/deformation

Applying mechanical stretching or deforming a foamed polymer can trigger strain-induced crystallization, especially in materials like polylactic acid (PLA). This process speeds up crystallization, leading to more chain orientation and crystallinity. These changes can, in turn, significantly affect the foam's mechanical properties and potentially its interconnectivity. Specifically, as cells grow, their walls experience biaxial stretching, while the struts (the thicker parts between cells) undergo uniaxial stretching.

These post-foaming treatments offer a second opportunity to modify the cellular structure, especially interconnectivity. This is particularly valuable for applications such as in biomedical scaffolds or filtration membranes. These methods enable modifications that might be difficult or impossible to achieve solely through primary foaming parameters, effectively decoupling some aspects of cell formation from cell interconnectivity.

2.3.9 Polymer blending and chemical modification for interconnectivity

Beyond process and post-foaming treatments, the inherent design of the polymer material itself, through polymer blending and chemical modification, offers proactive strategies for controlling cell interconnectivity.

2.3.10 Blending immiscible polymers

This method involves blending two or more polymers that do not easily mix, creating a heterogeneous melt structure. The boundaries between these

DOI: http://dx.doi.org/10.5772/intechopen.1012450

unmixable phases act as nucleation sites for gas bubbles. As these bubbles grow and collide at these interfaces, they thin and eventually rupture the cell walls, leading to open-cell structures. For example, when creating foams from PLA and poly(butylene succinate) (PBS), researchers have achieved a very high open-cell content (up to 97.7%), with PBS effectively acting as a cell opener [27]. This is a clever technique that uses phase separation to precisely control the foam's final shape and structure.

2.3.11 Chemical modifications

Setting a polymer's molecular architecture is a direct way to control its rheological properties and, by extension, a foam's interconnectivity. Techniques like chain extension, branching, and crosslinking are used to boost the polymer's melt strength. A stronger melt is vital; it prevents cells from collapsing or rupturing, keeping the cell walls intact. These modifications typically ramp up the polymer's molecular weight and viscosity, which helps reduce the loss of blowing agent gas during expansion. Plus, these chemical changes can form crystalline regions, acting as extra spots for bubbles to form, leading to a finer foam structure.

2.3.12 Advanced plasticizers

Certain specialized plasticizers [28] can be incorporated to enhance pore interconnectivity. For example, eugenol has been shown to improve pore interconnectivity in PCL foams, while also offering bioactive properties. When it comes to designing materials for specific foam structures, a smart strategy involves both blending different materials and modifying them chemically.

One clever way to manage how cells connect within a foam is by using immiscible blends. These blends, where the materials do not mix, naturally separate into distinct phases. This phase separation can be precisely controlled to intentionally rupture cell walls, creating the desired foam structure.

Another critical aspect is directly improving the melt strength of the material. A stronger melt means the cell walls can maintain their integrity better during the foaming process, which is essential for controlling the final foam structure.

3. Functionalization *via* supercritical impregnation

Material functionalization is a key aspect of advanced materials engineering, enabling existing structures to take on new properties for specific applications. In this field, supercritical impregnation (SSI) is a cutting-edge technique for imparting specific functionalities to polymeric foams. This section will provide a detailed discussion of the principles and mechanisms governing SSI, the critical factors influencing its effectiveness and a comprehensive review of its most recent and promising applications in various technological fields.

3.1 Principles and mechanisms of impregnation with $scCO_2$ in polymers

Supercritical solvent impregnation (SSI) is a modern technique that uses the unique transport properties of supercritical fluids to fabricate materials with advanced functionalities. The most commonly employed supercritical fluid is $scCO_2$, which exhibits an ideal combination of low viscosity, high diffusion coefficients, liquid-like densities, and virtually zero surface tension. These characteristics are critical to the success of the process [29].

The SSI mechanism proceeds in several interconnected stages. First, the $scCO_2$ acts as an effective solvent for the desired active ingredient, whether a drug, dye, bioactive compound, or nanoparticle. Thanks to its high diffusivity and zero surface tension, the $scCO_2$ solution and the active ingredient penetrate deep into the polymer matrix, even into the most intricate pores and microstructures of the material [29]. This deep, homogeneous penetration capability is a distinct advantage of SSI over conventional impregnation techniques, which often only reach the surface of the material or result in non-uniform distribution of the active agent.

Once the active ingredient has been dissolved and distributed evenly throughout the polymeric material, the system proceeds to the depressurization stage. During this critical phase, the CO_2 rapidly transitions from a supercritical state to a gaseous state and leaves the system. As a result of this controlled depressurization, the active principle precipitates and is uniformly deposited in the polymeric matrix. The final product is a dry, impregnated material free of organic solvent residues, simplifying further processing and purification steps. SSI can be implemented in batch mode for preliminary and laboratory-scale studies, and in semi-continuous mode for greater efficiency and scalability [29].

The effectiveness of impregnation is intrinsically linked to the complex interactions established between $scCO_2$, the polymer, and the active compound [30]. The $scCO_2$ acts as a solvent for the active compound and interacts directly with the polymer matrix. When absorbed by the polymer, the $scCO_2$ acts as a plasticizer, increasing the polymer's fractional free volume and reducing its intermolecular interaction energy [14]. This decreases the polymer's Tg and viscosity, thus facilitating the active compound's diffusion through the matrix [8]. The high diffusivity of the $scCO_2$ allows the active compound to penetrate faster and deeper into the polymer volume, ensuring a homogeneous distribution that is difficult to achieve with conventional methods.

Depressurization is not merely the removal of the solvent; it is a critical event that induces the precipitation of the solute and the 'fixation' of the new foam morphology. This interaction involves a complex balance of solute solubility in $scCO_2$, polymer plasticization and depressurization kinetics. The success of SSI hinges on the precise manipulation of these interface dynamics. Controlling pressure and temperature affects not only solubility, but also the degree of plasticization and the rate of solute diffusion. This makes SSI a 'real-time materials engineering' technique, where the fluid-polymer interaction is controlled to produce specific properties. The impregnation technique can be applied in a single step while the polymer is being plasticized and the compound to be impregnated is introduced into the mold, causing subsequent expansion when the pressure is released. **Figure 1** shows a diagram of the single-step process, which has been researched in several scientific publications [31].

Despite its many advantages, $scCO_2$ has one fundamental limitation: its poor solubilization capacity for polar and ionic species. This restricts the range of compounds that can be efficiently impregnated. However, this limitation has driven innovation in the field. To overcome it, research has explored the use of "binding agents" or "co-solvents" such as ethanol, water, or acetone [32]. These co-solvents can enhance the solubility of CO_2 in the polymer and increase the interaction between CO_2 and polymer chains, allowing impregnation of more polar compounds and modification of the porous structure. This inherent limitation of $scCO_2$ is not a dead end, but a catalyst for innovation, significantly broadening the spectrum of materials and composites that

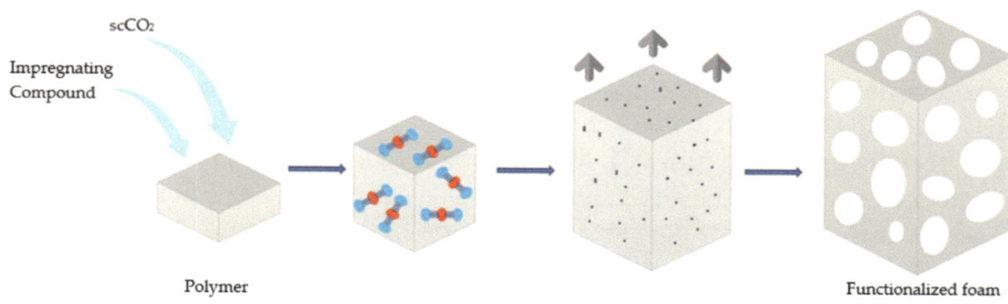

Figure 1.
Scheme of polymer foam formation and impregnation from CO_2 in a single-step.

can be functionalized, albeit at the cost of increased process complexity. Optimizing the SSI process requires a thorough understanding of the key factors that influence its development. **Table 1** summarizes these factors and their impact on the impregnation of polymeric foams with $scCO_2$.

3.2 Advanced applications and recent case studies

The ability to functionalize polymeric foams through impregnation with $scCO_2$ has opened up a wide range of advanced applications, demonstrating the versatility of this technology across various sectors. The impact of this technique is illustrated by recent case studies presented below.

3.2.1 Catalytic functionalization and nanoparticles

Polymeric foams are ideal supports for catalysts and nanoparticles thanks to their high surface area and controllable porous structure. Using $scCO_2$ enables the

Key factor	Summary of influence
Pressure	Controls $scCO_2$ density and solvent power; higher pressure increases CO_2 solubility and affects foam structure.
Temperature	Affects solute solubility and polymer-$scCO_2$ diffusion; also influences polymer crystallinity and thermal properties.
Impregnation time	Determines how much and how well the active compound is loaded and distributed; overly long times may reduce effectiveness.
Depressurization rate	Key for proper solute precipitation; too fast can cause uneven distribution or unwanted foaming.
Nature of the active compound	Solubility in $scCO_2$ depends on polarity and molecular structure; nonpolar or specific structures dissolve better.
Nature of the polymer	Polymer-$scCO_2$ affinity and crystallinity affect CO_2 absorption and solute diffusion; polymer plasticization is crucial.
Use of Co-solvents/binding agents	Helps dissolve polar compounds and enhance interactions, but can complicate the process.
Compound-polymer affinity	Strong interactions are essential for retaining and evenly distributing the active compound in the matrix.

Table 1.
Key factors influencing $scCO_2$ impregnation in polymers [6].

production of nanocellular foams with a diameter of less than 100 nm, which significantly improves catalytic activity by increasing the surface area. Foams stabilized with nanoparticles (e.g., nanosilica) have been developed for enhanced oil recovery (EOR), achieving stable structures containing 100–150 nm particles. Inorganic TiO_2 has also been incorporated into PCL foams using $scCO_2$-assisted foaming to obtain composite materials with density values in the 0.2–0.5 g/cm^3 range, pore size between 360 and 720 nm and narrow pore size distribution [33]. Furthermore, coating metal catalysts (Ag, Au, Pt, Pd, Zn and Cu) with polymers has been shown to improve activity, stability, and selectivity. For instance, biopolymers such as chitosan have been employed to coat catalytic metal-organic frameworks (MOFs) [34]. Finally, the $scCO_2$ technique has been shown to enable the tuning of pore size and density in PMMA foams, making it useful for designing customized catalytic supports.

Growing concerns about public health and food safety have led to the development of polymeric foams with antimicrobial properties. Impregnation with supercritical carbon dioxide is an ideal, nontoxic technique for biomedical applications, as it avoids the harmful residues left by conventional, solvent-based methods.

3.2.2 Antimicrobial functionalization

Recently, research has focused on impregnating polymers with natural bioactive compounds and plant extracts to impart antimicrobial activity. For instance, polypropylene nonwoven fibers impregnated with thymol have demonstrated high efficacy against common bacteria and fungi, such as *Escherichia coli*, *Staphylococcus aureus,* and *Candida albicans*, with loading reaching up to 19.6%. Similarly, 15% carvacrol was incorporated into cotton gauze, yielding comparable antibacterial results [29].

In tissue engineering, PCL and PCL-HA scaffolds impregnated with carvacrol were fabricated, reaching a loading of up to 10.57% in PCL-HA and demonstrating the potential of this approach for functionalized biomaterials [35]. Additionally, clove extract impregnated into polycarbonate (PC) achieved a loading of up to 18.5%, suggesting its potential use in antimicrobial food packaging thanks to the presence of eugenol [36]. These cases confirm that solid impregnation with $scCO_2$, also known as supercritical fluid impregnation (SFI), allows the development of safe, functional and antimicrobial materials that can be adapted for various applications.

3.2.3 Drug and bioactive compound delivery systems

The development of controlled drug release systems has been transformed by $scCO_2$ impregnation, which allows polymeric nanocarriers to be formulated with precise control of particle size, morphology, and drug loading. Furthermore, it eliminates the need for the purification steps that are common in traditional techniques [37]. For instance, impregnating aspirin in poly(lactic acid) (PLA) and linear low-density polyethylene (LLDPE) yielded higher loadings and a more prolonged release than traditional methods. PLLA released 60% of the drug within 74 days, which is ideal for long-term treatment. Similarly, the release of thymol from TPS and TPS-PCL mixtures can be modulated by varying the temperature, time, and PCL content [38]. Sustained release for up to 5 days has also been achieved under specific pressure, temperature and depressurization conditions [39]. Furthermore, micronation techniques using $scCO_2$ improve oral solubility and bioavailability, thereby reducing dosage and side effects. This also facilitates the development of pulmonary delivery systems [12].

Another example is the impregnation of PHB-HV with mango leaf extract to create materials with antioxidant and antimicrobial properties, which are modulated by operating conditions.

3.2.4 Fabric engineering scaffolds

Porous polymeric scaffolds are essential in tissue engineering, as they offer a three-dimensional structure that allows cell adhesion, proliferation and regeneration. The use of $scCO_2$ has established itself as a "green" alternative by avoiding organic solvents and high temperatures that could damage cells or sensitive molecules [32]. PCL scaffolds have been obtained by $scCO_2$-assisted foaming, achieving pores of 70–180 μm, 90% porosity and 96% interconnectivity, ideal for regenerative applications [11]. PDLLA and PLGA scaffolds, capable of releasing growth factors in a controlled manner, have also been fabricated, showing the versatility of the method.

An outstanding advance is the combination of 3D printing with supercritical foaming, which allows obtaining anisotropic structures with enhanced interconnected porosity, overcoming the limitations of conventional 3D printing. More homogeneous structures have also been achieved in crystalline polymers such as PCL, using $scCO_2$-ethanol blends [40].

3.3 Advantages and challenges of functionalization of polymeric foams with $scCO_2$

The functionalization of polymeric foams with $scCO_2$ is a highly efficient process intensification strategy, as it often allows foaming and functional loading to occur in a single step. This reduces the need for intermediate steps, as well as energy consumption and waste generation. The main advantages of this method are its sustainability and solvent-free nature, as it uses recycled CO_2 as the process medium, thereby decreasing the environmental footprint. The mild operating conditions (low temperature and inert atmosphere) enable the incorporation of thermo- or photo-sensitive compounds without causing degradation. Furthermore, the ability to adjust pressure and temperature provides precise control over pore morphology, density, and distribution, as well as the loading and release of functional agents. This unified approach clearly demonstrates intensification, as it reduces the number of steps, improves mass transport efficiency, and increases overall process efficiency. The result is purer, more active products with custom-designed structures from a single integrated operation.

However, there are challenges that hinder its industrial adoption. The high cost of pressurized equipment makes it difficult to scale up. Additionally, the low solubility of polar compounds in $scCO_2$ limits its scope without the use of co-solvents, which partly compromises the 'green chemistry' of the process. Morphological control scaling and achieving homogenous functional loading remain technical challenges, as does the end-of-life management of treated foams. Furthermore, the interaction of CO_2 with certain polymers can induce undesirable phenomena, such as secondary foaming or structural alterations, necessitating precise control of the operating conditions [41].

In conclusion, the functionalization of foams with $scCO_2$ embraces the spirit of process intensification, offering higher efficiency, a lower environmental impact, and multifunctional capabilities. Overcoming technical and economic barriers will be key to consolidating its large-scale industrial adoption.

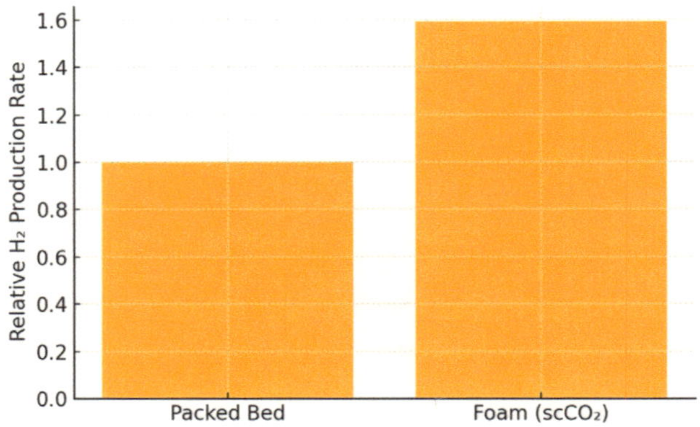

Figure 2.
Comparison of relative hydrogen production rate between packed bed and $scCO_2$-processed foam catalyst under identical conditions: dehydrogenation of H_{18} DBT at a temperature between 280 and 320°C at atmospheric pressure [46].

4. Applications in-process intensification

4.1 Polymeric foams as catalytic supports: Performance and process intensification potential

In the field of catalysis, open-cell foams are exceptional candidates for supporting structured catalysts in gas-solid catalytic processes. Their design enables them to achieve high reaction rates controlled by gas-solid diffusional mass transfer. These foams offer superior heat and mass transfer properties combined with low pressure drops, outperforming traditional packed beds in terms of high porosity volume [42]. Specific examples include palladium-alumina-coated metal foams for catalytic CO oxidation and metal supports in the form of honeycomb monoliths and open-cell foams as strong candidates for intensifying nonadiabatic catalytic processes [43]. In addition to catalysis, the controllable porosity and tunable surface functionality of porous polymers make them ideal for use in applications that require enhanced mass transfer, such as air purification and water treatment [44].

Recent studies have demonstrated that Pd/HDPE foams formed using $scCO_2$ exhibit open-cell structures (20–60 μm), significantly outperforming pellet-based supports in hydrogenation reactions [45]. Musavuli et al. (2025) discovered that Pt/Al_2O_3 catalysts supported on metal foams yielded 59% more hydrogen in LOHC hydrogenation processes than conventional packed beds (**Figure 2**) [46].

Jiang et al. demonstrated that $scCO_2$-assisted synthesis of metal nanoparticles enables precise control of morphology and loading, which is essential for upgrading biomass-derived molecules [47]. Furthermore, the oxidation of Cu nanoparticles in $scCO_2$, followed by electroreduction, generated amorphous CuO structures with over 65% Faradaic efficiency for C_2^+ products [48]. Together, these findings suggest the strong potential for integrating functionalized foams into compact reactor modules.

4.2 Advance media for mass and heat transfer

The physical properties of $scCO_2$—liquid-like density, gas-like diffusivity, and low viscosity—(**Figure 3**) enhance its role as a mass and heat transfer agent. Significant improvements in-process efficiency are the result of combining these properties with porous foams.

As illustrated in **Figure 3**, supercritical CO_2 has a density similar to that of a liquid and a diffusivity and viscosity similar to those of a gas. This combination offers unique advantages for mass and heat transfer. It is important to note that the density of typical process gases at ambient pressure and temperature is several orders of magnitude lower than that of supercritical CO_2 and liquids, often being less than 1 kg/m^3.

Versteeg et al. developed a model to analyze the correlations between the Sherwood and Reynolds numbers in $scCO_2$ flows [49]. These correlations are of fundamental importance for the design of heat exchangers and foam reactors. Furthermore, Nguyen et al. employed direct numerical simulations to unveil orientation-dependent heat transport behavior close to the critical point [50].

Experimental studies involving porous silicon carbide (SiC) tubes exposed to $scCO_2$ have confirmed superior convective heat transfer, thus indicating that foam-integrated $scCO_2$ [51]. Furthermore, Coupled-Matrix-Foam simulations enable the prediction of multiphase behavior in real-time reactor operation [52].

4.3 Lightweight and thermally insulative materials

$ScCO_2$-foamed polymers, including polylactic acid (PLA) and polycaprolactone (PCL), have been demonstrated to yield ultralight structures with low thermal conductivity, a property that is paramount for the insulation of reactors. The utilization of these materials as diffusion barriers or passive structural components is a subject that is being investigated with increasing frequency. Zhou et al. reviewed hybrid processing strategies (e.g., freeze-drying combined with supercritical carbon dioxide foaming) to generate uniform microcellular foams [53]. Furthermore, TPU-based foams processed using $scCO_2$ offer flexibility and durability, rendering them ideal for adaptive insulation in dynamic environments [54].

4.4 Integration into foam reactors and diffusion layers

Functional foams created with $scCO_2$ are increasingly incorporated into integrated reactor modules, where reaction, separation, and heat exchange co-occur. $scCO_2$ enables in-situ nanoparticle synthesis and impregnation, reducing steps and waste.

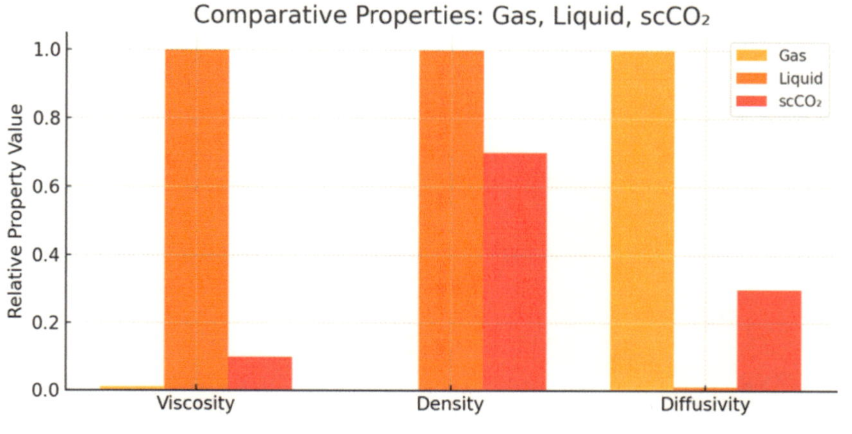

Figure 3.
Relative comparison of gas, liquid, and $scCO_2$ properties. Source: Own elaboration based on data from Ref. Brunner [35].

Sauceau et al. showed the new challenges, as a one-step $scCO_2$ extrusion method for catalytic foams, eliminating the need for post-processing [7]. Jiang et al. and Nguyen et al. both highlight how these foams are used in biomass valorization, gas reforming, and antimicrobial layer formation [47, 50]. The potential to use these foams in electrochemical systems—where $scCO_2$-processed Cu-based catalysts show enhanced C_2 product selectivity—demonstrates their versatility and alignment with green chemistry goals [48].

5. Environmental, energy and recycling benefits and scale-up challenges

5.1 Environmental sustainability and energy efficiency

The utilization of $scCO_2$ obviates the requirement for toxic chemical blowing agents. In contrast to chemical foaming agents, which release harmful by-products such as azodicarbonamide decomposing to CN compounds, $scCO_2$ is nontoxic, inert, and abundant, aligning with the principles of green chemistry [9].

The utilization of $scCO_2$ foaming as a substitute for synthetic blowing agents, in conjunction with a reduction in solvent usage, results in a substantial decrease in CO_2 and VOC emissions. Evaluations suggest that it mitigates the impact of greenhouse gases more effectively than conventional polymer foam production.

Integration of $scCO_2$ foaming with extrusion, microwave-assisted heating, or ultrasound-assisted batch foaming has been demonstrated to reduce energy demand (**Figure 4**) by up to 45% in comparison with traditional methods [52]. The application of ultrasound in conjunction with $scCO_2$ foaming has been demonstrated to enhance the porosity of the resultant material, whilst concomitantly reducing both the temperature and duration of the process.

Since $scCO_2$ does not introduce residual solvents and uses physical desorption, polymer foams can be reprocessed more easily. Initial studies demonstrate the feasibility of recovering and re-foaming PCL, PLA, and PMMA in closed-loop systems, thereby supporting circular economy models [55].

5.2 Technical and scale-up challenges

The consistent production of cellular morphology in continuous processes remains challenging, particularly for polymers such as polypropylene, which possess low melt

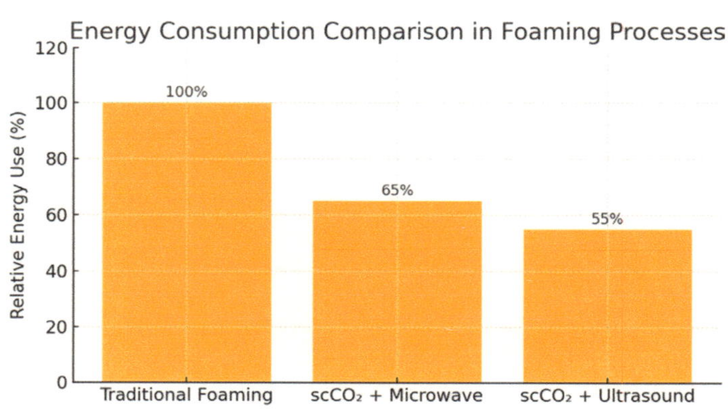

Figure 4.
Relative energy consumption in foaming processes using traditional vs. $scCO_2$-assisted methods [9, 52].

Benefit/challenge	Highlights
Environmental	Solvent-free, reduced GHG emissions, supports circular economy
Energy efficiency	Integrated heating/foaming methods reduce power demand up to ~45%
Morphological control	Requires rheological modifiers and precise process variables
Capital expense	High equipment costs limit adoption at mid-scale
Foam stability and durability	Needs robust design against mechanical and environmental degradation
Recycling infrastructure	Recycling pathways still underdeveloped
Limited agent diversity	Predominantly CO_2, few viable alternatives studied

Table 2.
Summary of key points.

strength. It is evident that techniques such as long-chain branching, chain extenders, blending, or fillers are utilized in order to enhance rheology. However, it is imperative to note that these techniques require meticulous fine-tuning [56].

The financial outlay required at the outset for $scCO_2$ foaming systems, including those of the Mucell® brand, and for custom extrusion units, is considerable. According to industry reports, capital costs are estimated to exceed those of traditional foam lines by approximately 30–40%, thereby reducing economic appeal at moderate scales [9].

The stabilization of foam under real-world conditions is contingent upon precise control of gas diffusion, polymer crystallinity, additives, and saturation/desaturation kinetics. In the absence of optimization, the integrity of the foam may be compromised, resulting in premature loss of its designated properties.

Despite the chemical superiority of $scCO_2$ foaming, the recycling of foamed polymer waste is not yet subject to a standardized set of methodologies. In the absence of effective reclamation technologies, the environmental benefits of these materials may be compromised.

The predominant agents in use at present are $scCO_2$ and, to a lesser extent, scN_2. Exploration of alternative supercritical gases remains in its infancy and is confronted with challenges pertaining to leakage, cost, and safety (**Table 2**).

6. Challenges and future perspectives

6.1 Future perspectives and hybrid composite foams

In the context of increasing demand for multifunctional, sustainable materials, there has been a notable integration of $scCO_2$ foaming with advanced processing techniques, leading to the production of hybrid foams. The combination of additive manufacturing (3D printing) and $scCO_2$ enables hierarchical architectures with tunable properties suitable for thermal management, shock absorption, and filtration.

Recent advancements in the field have focused on the utilization of nanocomposites comprising carbon nanotubes, graphene oxide, and lignin-derived particles, resulting in enhanced electrical conductivity, mechanical robustness, and thermal stability. These hybrid foams have also been demonstrated to be attractive for electromagnetic interference shielding and flexible electronics [56].

Emerging area	Key $scCO_2$ foam advantage
Hybrid nanocomposites	Enhanced conductivity, strength, and thermal stability
3D printing integration	Custom architectures, complex porous designs
EOR and CO_2 sequestration	Foam stability, shear thinning, and conformance control
Smart sensors and soft actuators	Flexible, resilient, and pressure-sensitive foams
Building insulation and packaging	Lightweight, low thermal conductivity, and green processing
Biomedical scaffolds	Porous, biocompatible, and antimicrobial surface functionalization

Table 3.
Summary of trends and applications.

The introduction of bio-based components (for example, PLA with lignin or cellulose nanofibers) into $scCO_2$ foamed structures is an area of research that is focused on reducing the environmental impact of the process, whilst maintaining or improving performance characteristics. These directions align with the overarching objective of the circular economy and are currently under active exploration.

6.2 Emerging applications in intensified systems

$scCO_2$-processed foams are being explored in various cutting-edge applications, especially in chemical engineering and energy systems:

- In enhanced oil recovery (EOR), $scCO_2$ foams exhibit shear-thinning behavior and improved reservoir conformance, enhancing sweep efficiency and reducing gas mobility. This supports CO_2 sequestration and recovery in fractured formations [57].

- In the domain of soft robotics and sensors, TPU-based $scCO_2$ foams with conductive fillers are utilized as pressure sensors and actuators, benefiting from their compressibility, resilience, and functional surface area.

- Advanced insulation panels for sustainable buildings and high-efficiency packaging leverage the low thermal conductivity and lightweight structure of $scCO_2$ foams. These panels often combine aerogels or phase-change materials within a foam matrix for energy regulation [10].

- Additionally, researchers are evaluating the use of foamed materials in biomedical applications such as drug delivery scaffolds and antimicrobial surfaces, particularly for implantable systems where porosity and sterility are crucial (**Table 3**) [58].

7. Conclusions

The use of $scCO_2$ as a foaming and impregnation agent represents an effective and sustainable strategy for the design of porous polymeric materials with advanced functionalities. Throughout this chapter, we have demonstrated how the precise tuning of parameters such as pressure and temperature allows for the control of polymer morphology—pore size, density, and degree of interconnectivity—while facilitating the incorporation of active compounds in a single processing step.

The cases analyzed include catalytic supports with improvements of up to 60% in hydrogen production, scaffolds with interconnected porosities greater than 90% and controlled release systems with profiles sustained for several days. These results confirm the potential of $scCO_2$-processed foams in applications aimed at process intensification, especially in catalysis, biomedicine, and heat and mass transfer.

While the environmental and functional advantages of using $scCO_2$ are evident—solvent-free processes, energy reduction, and recyclability—challenges remain related to scalability, compatibility with polar compounds, and end-of-life management of materials. Overall, polymer foams obtained using $scCO_2$ are a versatile tool for the development of compact, efficient, and environmentally friendly systems, in line with the principles of the circular economy and emerging technologies in chemical engineering.

Acknowledgements

We gratefully acknowledge the Spanish Ministry of Science and Innovation (Project PID2020-116229RB-I00) for financial support.

Conflict of interest

The authors declare no conflict of interest.

Author details

Ignacio García-Casas*, Antonio Montes, Ludisbel León-Marcos, Clara Pereyra and Diego Valor*
Faculty of Sciences, Department of Chemical Engineering and Food Technology, Wine and Agrifood Research Institute (IVAGRO), International Excellence Agrifood Campus (CeiA3), University of Cadiz, Cadiz, Spain

*Address all correspondence to: ignacio.casas@uca.es and diego.valor@uca.es

References

[1] Adewole JK, Yeneneh AM, Khan MY, Uddin MJ. Editorial: Membrane engineering and process intensification, front. Chemical Engineer. 2023;**5**:11-13. DOI: 10.3389/fceng.2023.1360708

[2] Strube J, Ditz R, Kornecki M, Huter M, Schmidt A, Thiess H, et al. Process intensification in biologics manufacturing. Chemical Engineering and Processing: Process Intensification. 2018;**133**:278-293. DOI: 10.1016/J.CEP.2018.09.022

[3] Whitford WG(B). Single-use perfusion bioreactors support continuous biomanufacturing. Pharmaceutical Bioprocessing. 2015;**3**:75-93. DOI: 10.4155/pbp.14.58

[4] Mills NJ, Zhu HX. The high strain compression of closed-cell polymer foams. Journal of the Mechanics and Physics of Solids. 1999;**47**:669-695. DOI: 10.1016/S0022-5096(98)00007-6

[5] Kulshreshtha A, Dhakad SK. Preparation of metal foam by different methods: A review. Materials Today Proceedings. 2020;**26**:1784-1790. DOI: 10.1016/J.MATPR.2020.02.375

[6] Peach J, Eastoe J. Supercritical carbon dioxide: A solvent like no other. Beilstein Journal of Organic Chemistry. 2014;**10**:1878-1895. DOI: 10.3762/bjoc.10.196

[7] Sauceau M, Fages J, Common A, Nikitine C, Rodier E. New challenges in polymer foaming: A review of extrusion processes assisted by supercritical carbon dioxide. Progress in Polymer Science. 2011;**36**:749-766. DOI: 10.1016/J.PROGPOLYMSCI.2010.12.004

[8] Razzak FSA. Review on green solvent supercritical carbon dioxide and its chemical reactions. Journal of Scientific Research. 2021;**65**:17-22. DOI: 10.37398/jsr.2021.650804

[9] Zhou Y, Tian Y, Peng X. Applications and challenges of supercritical foaming technology. Polymers (Basel). 2023;**15**:1-16. DOI: 10.3390/polym15020402

[10] Zhang Y, Shi L, Li J, Chen S, Yi Q. Recent advances in supercritical carbon dioxide foaming Technology for Thermoplastic Polyurethane Foams: Process optimization, property modulation, and emerging applications. Journal of Polymer Science. 2025;**63**:1-18. DOI: 10.1002/pol.20250171

[11] Song C, Luo Y, Liu Y, Li S, Xi Z, Zhao L, et al. Fabrication of PCL scaffolds by supercritical CO_2 foaming based on the combined effects of rheological and crystallization properties. Polymers (Basel). 2020;**12**:1-14. DOI: 10.3390/POLYM12040780

[12] Ruiz HK, Serrano DR, Calvo L, Cabañas A. Current treatments for COVID-19: Application of supercritical fluids in the manufacturing of Oral and pulmonary formulations. Pharmaceutics. 2022;**14**. DOI: 10.3390/pharmaceutics14112380

[13] Torres A, Romero J, Macan A, Guarda A, Galotto MJ. Near critical and supercritical impregnation and kinetic release of thymol in LLDPE films used for food packaging. Journal of Supercritical Fluids. 2014;**85**:41-48. DOI: 10.1016/J.SUPFLU.2013.10.011

[14] Gromov OI, Kostenko MO, Petrunin AV, Popova AA, Parenago OO, Minaev NV, et al. Solute diffusion into polymer swollen by supercritical CO_2 by high-pressure electron

paramagnetic resonance spectroscopy and chromatography. Polymers (Basel). 2021;**13**. DOI: 10.3390/polym13183059

[15] Kikic I, Vecchione F, Alessi P, Cortesi A, Eva F. Polymer plasticization using supercritical carbon dioxide: Experiment and modeling. Industrial & Engineering Chemistry Research. 2003;**42**:3022-3029

[16] Jacobs LJM, Kemmere MF, Keurentjes JTF. Sustainable polymer foaming using high pressure carbon dioxide: A review on fundamentals, processes and applications. Green Chemistry. 2008;**10**:731-773. DOI: 10.1039/b801895b

[17] García-Casas I, Valor D, Elayoubi H, Montes A, Pereyra C. Morphological 3D analysis of PLGA/chitosan blend polymer scaffolds and their impregnation with olive pruning residues via supercritical CO_2. Polymers (Basel). 2024;**16**. DOI: 10.3390/polym16111451

[18] Mou W. Effect of shear factor on bubble nucleation of polystyrene using supercritical fluid as a foaming agent. Journal of Polymer Engineering. 2014;**34**:387-394. DOI: 10.1515/polyeng-2013-0126

[19] Xiaopeng Chen CAB, Feng JJ. Plasticization effects on bubble growth during polymer foaming. Polymer Engineering and Science. 2006;**46**:97-107. DOI: 10.1002/pen

[20] Valadez-gonzalez A. The effect of two commercial melt strength enhancer additives on the thermal, rheological and morphological properties of polylactide. Journal of Polymer Engineering. 2016;**36**:31-41. DOI: 10.1515/polyeng-2014-0322

[21] Jia G, Huang H, Niu J, Chen C, Weng J, Yu F, et al. Exploring the interconnectivity of biomimetic hierarchical porous Mg scaffolds for bone tissue engineering: Effects of pore size distribution on mechanical properties, degradation behavior and cell migration ability. Journal of Magnesium and Alloys. 2021;**9**:1954-1966. DOI: 10.1016/j.jma.2021.02.001

[22] León JM, Bernardo V, Rodríguez-pérez MÁ. Two-stage depressurization in gas dissolution foaming: The production of nanocellular materials free of defects. Macromolecular Materials and Engineering. 2020;**305**:1-10. DOI: 10.1002/mame.202000283

[23] Shutov FA, Henrici-Olivé G, Olivé S. 6 injection molding: Gas counter pressure process. In: Integral/Structural Polymer Foams. Berlin, Heidelberg: Springer; 1986

[24] Villamil Jiménez JA, Le Moigne N, Bénézet JC, Sauceau M, Sescousse R, Fages J. Foaming of PLA composites by supercritical fluid-assisted processes: A review. Molecules. 2020;**25**. DOI: 10.3390/molecules25153408

[25] Watson NJ, Johal RK, Glover Z, Reinwald Y, White LJ, Ghaemmaghami AM, et al. Post-processing of polymer foam tissue scaffolds with high power ultrasound: A route to increased pore interconnectivity, pore size and fl uid transport. Materials Science and Engineering: C. 2013;**33**:4825-4832. DOI: 10.1016/j.msec.2013.07.046

[26] Collins NJ, Bridson RH, Leeke GA, Grover LM. Particle seeding enhances interconnectivity in polymeric scaffolds foamed using supercritical CO_2. Acta Biomaterialia. 2010;**6**:1055-1060. DOI: 10.1016/J.ACTBIO.2009.08.013

[27] Yu P, Mi H, Huang A, Geng L, Chen B, Kuang T, et al. Effect of poly

(butylenes succinate) on poly (lactic acid) foaming behavior: Formation of open cell structure. Industrial & Engineering Chemistry Research. 2015;**54**:6199-6207. DOI: 10.1021/acs.iecr.5b00477

[28] Akrami H, Park CB. Advanced polycarbonate foam: Enhancing foamability and mechanical properties with toughened polycarbonate using nanofibrillation method and crosslinked rubbery EPDM networks. Journal of Manufacturing Processes. 2024;**127**:340-348. DOI: 10.1016/j.jmapro.2024.07.105

[29] Int Z, Med A, Zizovic I. Potential of supercritical solvent impregnation for development of materials with antibacterial Properties. International Archives of Medical Microbiology. 2017;**1**:1-6

[30] Ivanovic J, Knauer S, Fanovich A, Milovanovic S, Stamenic M, Jaeger P, et al. Supercritical CO_2 sorption kinetics and thymol impregnation of PCL and PCL-HA. Journal of Supercritical Fluids. 2016;**107**:486-498. DOI: 10.1016/J.SUPFLU.2015.07.001

[31] Satpayeva A, Rojas A, Tyrka M, Ksepko E, Galotto MJ, Zizovic I. Supercritical foaming and impregnation of Polycaprolactone and Polycaprolactone-hydroxyapatite composites with Carvacrol. PRO. 2022;**10**. DOI: 10.3390/pr10030482

[32] Bhamidipati M, Scurto AM, Detamore MS. The future of carbon dioxide for polymer processing in tissue engineering. Tissue Engineering Part B: Reviews. 2013;**19**:221-232. DOI: 10.1089/ten.teb.2012.0361

[33] Salerno A, Domingo C. Polycaprolactone foams prepared by supercritical CO_2 batch foaming of polymer/organic solvent solutions. Journal of Supercritical Fluids. 2019;**143**:146-156. DOI: 10.1016/j.supflu.2018.08.006

[34] Arya RK, Thapliyal D, Pandit A, Gora S, Banerjee C. Polymer coated functional catalysts for industrial applications. Polymers. 2023;**15**:2009

[35] Brunner G. Supercritical fluids: Technology and application to food processing. Journal of Food Engineering. 2005;**67**:21-33. DOI: 10.1016/J.JFOODENG.2004.05.060

[36] Martins A, Coutinho IT, Silva EK, Kato IT, Meireles MAA, Maia-obi LP, et al. Supercritical CO_2 impregnation of clove extract in polycarbonate: Effects of operational conditions on the loading and composition. PRO. 2022;**10**:2661

[37] Gangapurwala G, Vollrath A, De San Luis A, Schubert US. PLA/PLGA-based drug delivery systems produced with supercritical CO_2—A green future for particle formulation? Pharmaceutics. 2020;**12**:1-26. DOI: 10.3390/pharmaceutics12111118

[38] Properties SBC, Skoric ML, Milovanovic S, Zizovic I, Ortega-toro R, Santagata G, et al. Supercritical CO_2 impregnation of Thymol in thermoplastic release kinetics. Polymers (Basel). 2022;**14**(20):4360

[39] Valor D, Montes A, Monteiro M, García-Casas I, Pereyra C, de la Ossa EM. Determining the optimal conditions for the production by supercritical CO_2 of biodegradable plga foams for the controlled release of rutin as a medical treatment. Polymers (Basel). 2021;**13**. DOI: 10.3390/polym13101645

[40] Tsivintzelis I, Pavlidou E, Panayiotou C. Biodegradable polymer foams prepared with supercritical CO_2–ethanol mixtures as blowing

agents. Journal of Supercritical Fluids. 2007;**42**:265-272. DOI: 10.1016/J.SUPFLU.2007.02.009

[41] Mosquera E, Martini RE, Go ML, Ga A. CO_2-assisted impregnation/deposition of polymeric materials with pharmaceutical , nutraceutical, and biomedical applications: A review. The Journal of Supercritical Fluids. 2022;**191**:105763. DOI: 10.1016/j.supflu.2022.105763

[42] Giani L, Groppi G, Tronconi E. Mass-transfer characterization of metallic foams as supports for structured catalysts. Industrial and Engineering Chemistry Research. 2005;**44**:4993-5002. DOI: 10.1021/ie0490886

[43] Balzarotti R, Ambrosetti M, Beretta A, Groppi G, Tronconi E. Recent advances in the development of highly conductive structured supports for the intensification of non-adiabatic gas-solid catalytic processes: The methane steam reforming case study, front. Chemical Engineer. 2021;**3**:1-17. DOI: 10.3389/fceng.2021.811439

[44] Warren Z, Tasso Guaraldo T, Martins AS, Wenk J, Mattia D. Photocatalytic foams for water treatment: A systematic review and meta-analysis. Journal of Environmental Chemical Engineering. 2023;**11**:109238. DOI: 10.1016/J.JECE.2022.109238

[45] Liao W, Wu BZ, Chen HY, Yu JJ, Chiu KH. Fabrication of a form- and size-variable microcellular-polymer-stabilized metal nanocomposite using supercritical foaming and impregnation for catalytic hydrogenation. Nanoscale Research Letters. 2012;7. DOI: 10.1186/1556-276X-7-283

[46] Musavuli KC, Modisha P, Everson RC, Malakhov A, Bessarabov D. Metal foam as surface-extended catalyst support structure for process intensification in the dehydrogenation of Perhydro-Dibenzyltoluene on a Pt/Al_2O_3 catalyst. Catalysts. 2025;**15**. DOI: 10.3390/catal15010044

[47] Jiang Z, Zeng Y, Guo R, Lin L, Luque R, Yan K. Recent advances on CO_2-assisted synthesis of metal nanoparticles for the upgrading of biomass-derived compounds. Renewable and Sustainable Energy Reviews. 2024;**203**:114756. DOI: 10.1016/J.RSER.2024.114756

[48] Chen C, Yan X, Wu Y, Zhang X, Liu S, Zhang F, et al. Oxidation of metallic Cu by supercritical CO_2 and control synthesis of amorphous nano-metal catalysts for CO_2 electroreduction. Nature Communications. 2023;**14**. DOI: 10.1038/s41467-023-36721-8

[49] Versteeg FA, Picchioni F, Versteeg GF. On the mass transfer of supercritical fluids, specifically super critical CO_2: An overview. Chemical Engineering Journal. 2024;**493**:152521. DOI: 10.1016/J.CEJ.2024.152521

[50] Nguyen T, Merzari Ken E, Alice Lindquist M, Yuan H. Direct numerical simulation of high Prandtl number fluids and supercritical carbon dioxide canonical flows using the spectral element method. In: 19th International Meeting on Nuclear Reactor Thermohydraulics. arXiv. 2022

[51] Zhu T, Xiang M, Zhang L, Zhang S, Sun Z, Zhang J, et al. Design and manufacturing of Z-shaped silicon carbide printed circuit board heat exchanger. Journal of Supercritical Fluids. 2025;**225**:106709. DOI: 10.1016/J.SUPFLU.2025.106709

[52] Villamil Jiménez JA, Haurat M, Berriche R, Baillon F, Sauceau M, Chaussat M, et al. Polymer supercritical

CO_2 foaming under peculiar conditions: Laser and ultrasound implementation. Polymers (Basel). 2023;**15**. DOI: 10.3390/polym15081968

[53] Zhou Y, Tian Y, Zhang M. Technical development and application of supercritical CO_2 foaming technology in PCL foam production. Scientific Reports. 2024;**14**. DOI: 10.1038/s41598-024-57545-6

[54] Zhang C, Sun Z, Yang X, Duan G, Wang M, Qiao Y. Effect of thermoplastic starch content on the Properties of poly(butylene adipate-co-terephthalate) (PBAT) composites. BioResource. 2025;**20**:3910

[55] Lima GMR, Mukherjee A, Picchioni F, Bose RK. Characterization of biodegradable polymers for porous structure: Further steps toward sustainable plastics. Polymers (Basel). 2024;**16**. DOI: 10.3390/polym16081147

[56] Wang G, Dong M, Deng H, Ma X, Zhu B, Zhou L, et al. Polypropylene foaming using supercritical carbon dioxide: A review on fundamentals, technology, and applications. Advanced Composites and Hybrid Materials. 2025;**8**. DOI: 10.1007/s42114-024-01117-9

[57] Halsøy H, Graue A, Alcorn ZP. Supercritical CO_2 foam stability in the presence of oil. SSRN. n.d. Available from: https://ssrn.com/abstract=4990425

[58] Fei Y, Jiang R, Fang W, Liu T, Saeb MR, Hejna A, et al. Highly sensitive large strain cellulose/multiwalled carbon nanotubes (MWCNTs)/thermoplastic polyurethane (TPU) nanocomposite foams: From design to performance evaluation. Journal of Supercritical Fluids. 2022;**188**:105653. DOI: 10.1016/J.SUPFLU.2022.105653